The Innovation in Computing Companion

Gerard O'Regan

The Innovation in Computing Companion

A Compendium of Select, Pivotal Inventions

 Springer

Gerard O'Regan
SQC Consulting
Mallow, Cork, Ireland

ISBN 978-3-030-02618-9   ISBN 978-3-030-02619-6   (eBook)
https://doi.org/10.1007/978-3-030-02619-6

Library of Congress Control Number: 2018959840

This Springer imprint is published by the registered company Springer Nature Switzerland AG
The registered company address is: Gewerbestrasse 11, 6330 Cham, Switzerland

To
Joanne Di Natale (Jo)
In memory of a beautiful day in Taormina,
Sicily.

Preface

Overview

The objective of this book is to consider a selection of pivotal innovations that have shaped the computing field. The goal is to provide a brief account of each invention, as well as short biographical information on the inventor.

It is not feasible, due to space constraints, to consider all innovations that merit inclusion, and the selection inevitably reflects the bias of the author. It is hoped that that the reader will find the topics interesting and that she will gain an appreciation of some of the key inventions that have shaped the computing field.

Organization and Features

We discuss a selection of historical inventions such as Babbage's difference engine and his analytic engine. We present binary arithmetic, which was developed by Leibniz in the seventeenth century. We consider Boole's symbolic logic which provided the perfect model for the design of digital circuits and is the foundation of digital computing; and we discuss the von Neumann architecture which is the fundamental architecture underlying a computer.

We discuss a selection of historical analog and digital computers including Hollerith's tabulating machine, which was developed for census calculation in the late nineteenth century. Bush's differential analyser was developed for solving differential equations in the early twentieth century. Howard Aiken and IBM developed the Harvard Mark 1 analog computer, and it could execute long computations automatically. The Atanasoff-Berry Computer (ABC) was developed by John Atanasoff and Clifford Berry in the early 1940s to solve systems of linear equations. The Colossus computer was developed by Tommy Flowers and others for code-breaking work at Bletchley Park. The ENIAC and EDVAC computers were developed by John Mauchly and Presper Eckert in the mid-1940s in the United States.

The Manchester Mark 1 computer was developed at the University of Manchester in the late 1940s. The Z1–Z4 were developed by Konrad Zuse in Germany in the late 1930s/early 1940s. The LEO I computer was developed by Lyons in England in the early 1950s.

We discuss a selection of mainframes and minicomputers developed from the 1960s including the IBM System 360, the Amdahl 470 and 580 computers and the DEC PDP-11 and VAX-11/780 minicomputers.

We discuss a selection of programming languages including Ada, which was developed by the US Military. C was developed by Dennis Ritchie at Bell Labs in the early 1970s, and C++ was developed in the mid-1980s by Bjarne Stroustrup at Bell Labs. COBOL was developed by Grace Murray Hopper and the CODASYL committee in the late 1950s, and Java was developed in the mid-1990s by James Gosling at Sun Microsystems.

We discuss the invention of the transistor at Bell Labs, the invention of the integrated circuit at Texas Instruments and the invention of the microprocessor at Intel. We discuss a selection of operating systems including Unix and MS/DOS.

We discuss a selection of inventions in home and personal computers including the MITS Altair, the Apple I and II computers, the Commodore PET and 64 computers, the IBM personal computer and the Apple Macintosh. We discuss inventions related to personal computing including the mouse, the GUI, Atari video games and Microsoft Office.

We discuss several innovations related to fixed line and mobile communications including the AXE system, mobile phones, WiFi and Iridium. We discuss the Internet and World Wide Web, e-commerce, smartphones, social media and GPS.

We discuss several innovations in the software engineering field including the Agile methodology, CMMI, formal methods, object-oriented design and development, software inspections and software lifecycles.

We present a selection of inventions related to commercial computing including databases, open-source software development and cloud computing. Finally, we discuss miscellaneous inventions such as the ATM, AI, Eliza, digital photography, MP3 and digital music, robotics and Wikipedia.

The selection of innovations is presented in alphabetical order starting with the ABC computer and ending with Zuse's Z1–Z4 computers.

Audience

This book is suitable for computing students who are interested in knowing about inventions that have shaped the computing field. It will also be of interest to general readers who are curious about the many inventions in the computing field.

Cork, Ireland   Gerard O'Regan
2018

Acknowledgments

I am deeply indebted to friends and family who supported my efforts in this endeavour. My thanks to Joanne Di Natale (Jo) for friendship and special memories of Valencia, Nice and Rome. I would like to thank all copyright owners for permission to use their images. I believe that all necessary permissions have been obtained, but in the unlikely event that an image has been used without the appropriate authorization, please contact me so that the required permission may be obtained. Finally, I would like to thank the team at Springer for their constant professional work and support.

Contents

List of Figures

List of Tables

Chapter 1
Background

1.1 Introduction

Computers are an integral part of modern society, and new technology has transformed the world into a global village. Communication today is instantaneous and may be conducted using text messaging, email, mobile phones, and video calls over the Internet. In the past, communication involved writing letters, sending telegrams, or using home telephones. The revolutionary innovation in computer and information technology has allowed business to be conducted in a global market.

A computer is a programmable electronic device that can process, store, and retrieve data. The data is processed by a set of instructions termed a *program*, and all computers consist of two basic parts, namely, *hardware* and *software*. The hardware is the physical part of the machine, and a digital computer contains memory for short-term storage of data or instructions; a central processing unit for carrying out arithmetic and logical operations; a control unit responsible for the execution of computer instructions in memory; and peripherals that handle the input and output operations. The underlying architecture is referred to as the *Von Neumann architecture*, which is named after the mathematician, John von Neumann. Software is a set of instructions that tells the computer what to do and is created by one or more programmers. It differs from hardware in that it is intangible, whereas the hardware is physical.

The original meaning of the word *computer* referred to someone who carried out calculations rather than an actual machine. The early digital computers built in the 1940s and 1950s were enormous machines consisting of several thousand vacuum tubes[1]. They typically filled a large room or building, but their computational power was a fraction of the power of the computers used today.

---

[1] The Whirlwind Computer (developed in the early 1950s) occupied an entire building. One room contained the operating console consisting of several input and output devices.

© Springer Nature Switzerland AG 2018
G. O'Regan, *The Innovation in Computing Companion*,
https://doi.org/10.1007/978-3-030-02619-6_1

There are two distinct families of computing devices, namely, *digital computers* and the historical *analog computer*. These two types of computer operate on quite different principles, and the earliest computers were analog not digital.

The representation of data in an analog computer reflects the properties of the data that is being modeled. For example, data and numbers may be represented by physical quantities such as electric voltage in an analog computer, whereas in a digital computer, a stream of binary digits is used.

A digital computer is a sequential device that generally operates on the data one step at a time. The data in a digital computer are represented in binary format, which employs just two digits, namely, 0 and 1. A single transistor has two states, i.e., on and off, and is used to represent a single binary digit. Several transistors are used to represent larger numbers.

Early computing devices include the slide rule and various mechanical calculators. William Oughtred and others invented the slide rule in 1622, and it allowed multiplication and division to be carried out significantly faster than calculation by hand. Blaise Pascal invented the first mechanical calculator in 1642. It was called the Pascaline, and it could add or subtract two numbers. Multiplication or division could not be performed directly but could instead be carried out by repeated addition or subtraction.

Leibniz[2] invented a mechanical calculator (called the *Step Reckoner*) in 1672. This was the first calculator that could perform all four arithmetic operations: i.e., addition, subtraction, multiplication, and division.

James Thompson (who was the brother of Lord Kelvin) did early work on analog computation in the nineteenth century. He invented a wheel and disk integrator, which was used in mechanical analog devices, and he worked with Kelvin to construct a device to perform the integration of a product of two functions.

The operations in an analog computer are performed in parallel, and they are useful in simulating dynamic systems. They have been applied to flight simulation, nuclear power plants, and industrial chemical processes.

Vannevar Bush and others developed the first large-scale general-purpose mechanical analog computer at Massachusetts Institute of Technology in the late 1920s (Fig. 1.1). Bush's differential analyzer was designed to solve sixth order differential equations by integration, using wheel-and-disk mechanisms to perform the integration. The machine enabled integration and differential equation problems to be solved more rapidly.

The machine consisted of wheels, disks, shafts, and gears to perform the calculations, and it required significant effort to be set up by technicians to solve a particular problem. It contained 150 motors and miles of wires connecting relays and vacuum tubes.

Data representation in an analog computer is compact, but it may be subject to corruption with noise. A single capacitor can store one continuous variable in an analog computer, whereas several transistors are required to represent a variable in

---

[2]Leibniz is credited (along with Newton) with the development of the Calculus.

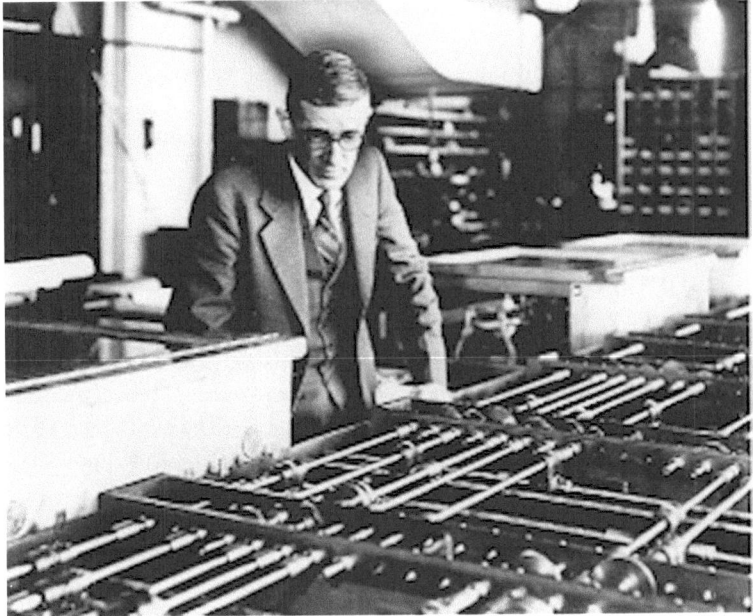

Fig. 1.1 Differential Analyzer at Moore School of Engineering, University of Pennsylvania

a digital computer. Analog computers were replaced by digital computers after the Second World War.

1.2   Digital Computers

Early digital computers used vacuum tubes to store binary information. A vacuum tube could represent the binary value "0" or "1." However, the tubes were large and bulky and generated a significant amount of heat. This led to problems with their reliability, and air-conditioning was employed to cool the machine.

Shockley and others at Bell Labs invented the transistor in the early 1950s, and they replaced vacuum tubes in the 1950s. Transistors are small and require very little power, and the resulting machines were smaller, faster, and more reliable.

Integrated circuits were introduced in the 1960s, and a massive amount of computational power could now be placed on a very small chip. Integrated circuits are small and consume a low amount of power, and they may be mass-produced to a very high-quality standard. Billions of transistors may be placed on an integrated circuit.

The development of the microprocessor allowed a single chip to contain all components of a computer from the CPU and memory to input and output controls. The microprocessor could fit into the palm of the hand, whereas the early computers filled an entire room.

The fundamental architecture of a computer has remained basically the same since Von Neumann and others proposed it in the 1940s. It includes a central processing unit, the control unit, the arithmetic-logic unit, an input and output unit, and memory.

1.3 Hardware and Software

Hardware is the physical part of the machine. It is tangible and may be seen and touched. It includes the historical punched cards and vacuum tubes, transistors and circuit boards, integrated circuits, and microprocessors. The hardware of a personal computer includes keyboard, network cards, mouse, DVD drive, hard disk drive, monitor, printers, scanners, and so on.

Software is intangible in that it is not physical as such, and instead it consists of a set of instructions that tells the computer what to do. It is an intellectual creation of a programmer or a team of programmers, and there are two main categories, namely, system software and application software.

The *system software* manages the hardware and resources and acts as an intermediary between the application programs and the computer hardware. This includes the UNIX operating system, the various Microsoft Windows operating systems, and the Mac operating system. There are also operating systems for mobile phones, video games, and so on. *Application software* is designed to perform a specific application such as banking, insurance, or accounting.

The early computer software consisted of instructions in machine code that could be immediately executed by the computer. These programs were difficult to read and debug, and this led to assembly languages where a mnemonic represented a machine code instruction. The assembly language program was translated into machine code by an assembler, and although the use of an assembly language was an improvement on directly entering machine code, it was still difficult to use. This motivated the development of high-level programming languages (e.g., Fortran and COBOL) where a program was written in the high-level language and compiled to the code of the target machine.

1.4 Innovations in Computing

The process of translating a business idea or invention into a product or service that adds value and that people will pay for is termed *innovation*. However, for a business idea to be termed innovative, it must be commercially viable at a cost that people will be willing to pay, and it must satisfy a specific customer need (as otherwise, there will be no demand for it).

There are two broad categories of innovations, namely, *evolutionary* and *revolutionary* innovations. Incremental advances in technology generally bring about an evolutionary innovation, whereas a revolutionary innovation is often totally new and completely different from the existing products in the marketplace (e.g., the development of the Apple Macintosh or iPhone was a paradigm shift from the existing state of the art). There is generally greater risk with a revolutionary innovation as it is creating an entirely new product, whereas evolutionary innovations generally involve less risk.

The success of hi-tech companies relies on the creativity and innovation of its staff, and it is therefore important to foster innovation in the workplace. An innovative work environment generally has a low power distance between management and staff, with an emphasis on open communication and inter-department collaboration. Brainstorming sessions to come up with innovative ideas or solutions to problems are encouraged, as well as the use of a suggestion box where employees can submit ideas or improvement suggestions as well as making them to their supervisor.

The objective of this book is to give a concise account of the work of a selection of innovations that have shaped the computing field, and the selection is presented in alphabetical order.

We discuss a selection of historical inventions such as Babbage's difference and analytic engines. The difference engine was designed to compute polynomial functions, and the Analytic Engine was essentially the design of the world's first mechanical computer. Boole's symbolic logic was recognized by Claude Shannon as providing the appropriate mathematical model for the design of digital circuits. Binary arithmetic was developed by Leibniz in the seventeenth century and is used extensively in digital computing, and the von Neumann architecture is the fundamental architecture underlying a computer.

We discuss a selection of historical analog and digital computers including Hollerith's tabulating machines; Bush's differential analyzer was the first large-scale general-purpose mechanical analog computer. The Harvard Mark 1 computer was an electromechanical calculator that was developed by Howard Aiken and IBM; the Atanasoff-Berry Computer (ABC) was invented by John Atanasoff in the early 1940s; the Colossus computer was developed by Tommy Flowers and others at Bletchley Park in England; the ENIAC and EDVAC computers were developed by John Mauchly and Presper Eckert in the United States; the Manchester Mark 1 computer was developed by Frederick Williams and others in England; the Z1–Z4 were developed by Konrad Zuse in Germany; and the LEO computers were developed by Lyons in England.

We discuss a selection of mainframes and minicomputers including the IBM System 360, which was designed by Gene Amdahl, the Amdahl 470 and 580 computers which were developed by the Amdahl Corporation, and the DEC PDP-11 and VAX-11/780 minicomputers.

We discuss a selection of programming languages including Ada, which was developed by the US military; C and C++ were developed at Bell Labs; COBOL was developed by Grace Murray Hopper and the CODASYL committee; and Java was developed at Sun Microsystems.

We discuss the invention of the transistor by William Shockley and others at Bell Labs. Jack Kilby at Texas Instruments invented the integrated circuit, and Tedd Hoff and others invented the microprocessor at Intel. We discuss the invention of operating systems such as Unix and MS/DOS.

We discuss a selection of inventions of home and personal computers including the MITS Altair, the Apple I and II computers, the Commodore PET and 64 computers, the IBM personal computer, and the Apple Macintosh. We discuss inventions related to personal computing including the mouse, the GUI, the Atari video games, and the Microsoft Office.

We discuss several inventions related to fixed line and mobile communications including the AXE system, mobile phones, Wi-Fi, and Iridium. We discuss the Internet and World Wide Web, e-commerce, smartphones and social media, and GPS.

We discuss several innovations in the software engineering field including the Agile methodology, CMMI, formal methods, object-oriented design and development, open-source software development, software inspections, and software lifecycles.

We present a selection of inventions related to commercial computing including databases and cloud computing. Finally, we discuss miscellaneous inventions such as the ATM, AI, Eliza, Digital Photography, MP3 and digital music, robotics, and Wikipedia.

The selection of innovations is presented in alphabetical order starting with the ABC computer and ending with Zuse's Z1–Z4 computers.

Chapter 2
ABC Computer

The Atanasoff-Berry Computer (ABC) was invented by John Atanasoff and built with the assistance of his graduate student, Clifford Berry (Fig. 2.1). Atanasoff was born in New York in 1903 and he earned a PhD in theoretical physics from the University of Wisconsin in 1930. He became an assistant professor at Iowa State College, where he taught mathematics and physics.

He became interested in developing faster methods of computation while doing his PhD research, as he wished to ease the time-consuming burden of calculation. He did some work on an analog calculator in 1936, but he concluded that analog devices were too restrictive and could not give the desired accuracy. His goal was to mechanize calculation to enable computation to be carried out faster.

The existing computing devices were mechanical, electromechanical, or analog. Atanasoff developed the concept of a digital machine to perform faster computation in the late 1930s, and he believed that his proposed machine offered advantages over the slower and less accurate analog machines. He published the design of a machine to solve linear equations using his own version of Gaussian elimination in the summer of 1939. He then used his research grant of $650 to build the ABC computer, with the assistance of his graduate student, Clifford Berry, from 1939 to 1942.

The ABC was approximately the size of a large desk and had approximately 270 vacuum tubes. Two hundred and ten tubes controlled the arithmetic unit; 30 tubes controlled the card reader and card punch; and the remaining tubes helped maintain charges in the condensers. It employed rotating drum memory with each of the 2 drum memory units able to hold 30 50-bit numbers.

The ABC was a digital machine that was designed for a specific purpose (i.e., solving linear equations) rather than as a general-purpose computer. The working prototype was one of the earliest electronic digital computers[1]. However, the ABC was slow and it required constant operator monitoring.

---

[1] The ABC was ruled to be the first electronic digital computer in the Sperry Rand vs. Honeywell patent case in 1973. However, it was preceded by Zuse's Z3 which was created in Germany in 1941.

© Springer Nature Switzerland AG 2018
G. O'Regan, *The Innovation in Computing Companion*,
https://doi.org/10.1007/978-3-030-02619-6_2

Fig. 2.1 Replica of ABC computer: creative commons

It used binary mathematics and Boolean logic to solve simultaneous linear equations. It employed over 270 vacuum tubes for digital computation, but it had no central processing unit (CPU) and was not programmable.

It weighed over 300 kg and used 1.6 km of wiring. Data were represented by 50-bit numbers. It performed 30 additions or subtractions per second. The memory and arithmetic units could operate and store 60 such numbers at a time ($60 \times 50 = 3000$ bits). The arithmetic logic unit was fully electronic and implemented with vacuum tubes.

The input was in decimal format with standard IBM 80 column punch cards, and the output was decimal via a front panel display. A paper card reader was used as an intermediate storage device to store the results of operations too large to be handled entirely within electronic memory. The ABC pioneered important elements in modern computing including:

- Binary arithmetic and Boolean logic.
- All calculations were performed using electronics rather than mechanical switches.
- Computation and memory were separated.

The ABC was tested and operational by 1942, and its historical significance is that the principles that it employed demonstrated the feasibility of electronic computing. Several of its concepts were later used in the ENIAC computer developed by Mauchly and Eckert[2].

---

[2] ENIAC is discussed in Chap. 23.

Atanasoff then commenced a Second World War assignment and worked for the US government in the postwar years. He received several honors including the US National Medal in Technology and Innovation, which was presented by President Bush in 1990.

The ABC computer was ruled to be the first electronic digital computer in the 1973 *Honeywell* vs. *Sperry Rand* patent court case in the United States. The court case arose from a patent dispute between Honeywell and Sperry-Rand[3], and John Atanasoff was called as an expert witness in the case.

John Mauchly (see Chap. 23) had visited Atanasoff and Berry in 1941, and they spent hours discussing the ABC computer and computer theory. Further, Mauchly viewed and observed the ABC computer in operation with Atanasoff and Berry for 3–4 days in 1941. He read Atanasoff's 35-page manuscript describing the construction and operation of the ABC, and he later asked Atanasoff whether he had any objections to him using some of the concepts used in the ABC in a machine that he was considering building.

The trial commenced in 1971 and sat for a total of 135 days, with over 70 witnesses presenting oral evidence and a further 80 witnesses providing written statements. The court made a final ruling in 1973 and declared that Eckert and Mauchly did not invent the first electronic computer, since the ABC existed as *prior art* at the time of their patent application.

It is fundamental in patent law that an invention is novel and that there is no existing prior art. This meant that the Mauchly and Eckert patent application for ENIAC was invalid, and the US court named John Atanasoff as the inventor of the first digital computer.

Controversy (Mauchly and Atanasoff)

Mauchly visited Atanasoff on several occasions and they discussed the implementation of the ABC computer. Mauchly subsequently designed the ENIAC, EDVAC, and UNIVAC computers. A 1973 legal case ruled that the Atanasoff-Berry Computer (ABC) existed as prior art at the time of the ENIAC patent application. The court ruled that the ABC was the first digital computer and stated that the inventors of ENIAC had derived the subject matter of the electronic digital computer from Atanasoff.

---

[3] Sperry-Rand held the patents for ENIAC (as they had been sold by Mauchly to Sperry-Rand).

Chapter 3
Ada Programming Language

The Ada programming language is named in honor of Lady Augusta Ada Lovelace (née Byron). Lovelace (Fig. 3.1) was an English mathematician who collaborated with Babbage (see Chap. 7) on applications of the Analytical Engine. She wrote what is considered the world's first program and is therefore considered the first programmer.

She was born in London in 1815 and was the only legitimate daughter[1] of the English poet, Lord Byron. Her parents separated shortly after her birth, and her mother, Annabella Milbanke, who was a mathematician, was granted sole custody of her. Lady Byron arranged for Ada to be educated in science and mathematics by private tutors. Lord Byron left England in 1816 never to return, and he died in Athens in 1824 when Ada was just nine. She never met her father but is buried beside him in the Church of St. Mary, Magdalene, Hucknall in Nottinghamshire.

She was presented at court in 1833 and introduced to Babbage at a dinner party later that year. She and her mother visited Babbage's studio in London, where the prototype Difference Engine was on display. She was fascinated by the machine and recognized its beauty.

She became friends with Mary Somerville (a British mathematician and astronomer), who guided her study of mathematics, and sent her mathematics books and problems to solve. They spoke regularly on mathematics and science, including Babbage's calculating machines. Ada was fascinated by the idea of the analytical engine, and she communicated regularly with Babbage with ideas on its applications.

---

[1] It is likely that Lord Byron was the father of Elizabeth Medora Leigh, as he is believed to have had an incestuous affair with his half sister, Augusta Leigh. Byron's former wife, Annabella Milbanke, told Ada that Leigh was her half sister and fathered by Byron. He was also the father of Clara Allegra Byron, the illegitimate daughter of Claire Clairmont (the stepsister of Mary Shelly). The child died of typhus aged five during Byron's travels in Italy.

© Springer Nature Switzerland AG 2018
G. O'Regan, *The Innovation in Computing Companion*,
https://doi.org/10.1007/978-3-030-02619-6_3

Fig. 3.1 Lady Ada
Lovelace

She married William King (who became the Earl of Lovelace in 1838) in 1835, and she became the Countess of Lovelace. She began studies in advanced mathematics with the logician, Augustus De Morgan, in 1841.

She produced an annotated translation of Menabrea's *Notions sur la machine analytique de Charles Babbage* (Ada, Augusta, Countess of Lovelace 1842). The notes that she added were about three times the length of the original paper and considered many difficult and abstract questions[2]. These notes are regarded as a description of a computer and software, and they were published in Richard Taylor's Scientific Memoirs (Vol. 3) in 1843.

She explained in the notes how the Analytical Engine could be programmed, and she wrote what is considered the *first computer program*. This program provided a written plan for how the Analytical Engine would calculate *Bernoulli numbers. She is therefore considered to be the first computer programmer*, and she was called the *enchantress of numbers* by Babbage.

The computer programming language developed in the late 1970s by the US Department of Defense was named *Ada* in her honor. The British Computer Society awards the *Lovelace medal* to individuals who have made an outstanding contribution to the understanding or advancement of computing, and the winner is invited to give the BCS public *Lovelace lecture* the following year. Her achievements remain an inspiration to women in science, engineering, and mathematics. There is an

---

[2]There is some controversy as to whether this was entirely her own work or a joint effort with Babbage.

annual "Ada Lovelace Day" held in October, and its goal is to raise the profile of women in mathematics, science, and engineering.

She suffered from health problems and had issues with alcohol addiction and gambling later in life. She died of cancer at the young age of 37 in 1852.

3.1   The Ada Language

The Higher Order Language Working Group (HOLWG) was formed in the mid-1970s with the goal of developing a programming language to replace (or reduce) the plethora of programming languages (several hundred) employed by the US Department of Defense (DoD) for its embedded system projects.

The HOLWG team defined the requirements that the new (or possibly existing) language was to satisfy, and the group formally reviewed existing programming languages to determine if any of them satisfied the specification. The team concluded in 1977 that none of the existing language met the requirements, and a request for a proposal to develop a new programming language was issued. Several proposals were received from organizations such as Honeywell, SRI, and Softech, and the Honeywell proposal was accepted.

The new language was given the name "Ada" after Lady Ada Lovelace. The preliminary Ada reference manual was published in 1979, and the Department of Defense military standard for the language was approved in 1980. It was given the classification number MIL-STD-1815 in honor of Ada (where 1815 is the year of her birth). The ANSI standard of the language (Ada 83) was published in 1983, and it became an ISO standard in 1987. An object-oriented version of the language (Ada 95) was published in 1995.

It was predicted that Ada would become the dominant general-purpose programming language, but it proved to be a challenge for the early compilers to implement the language as it was large and complex. The compilation time and run time performance was slow with the early compilers, and Hoare and others criticized Ada for being overly complex and unreliable. The earliest Ada implementation was the NYU Ada/Ed translator in 1983, and commercial compilers gradually became available leading to improvements in performance. The US Department of Defense began to mandate the use of Ada from 1991, and this continued up until 1997.

Ada provides support for safety-critical system, which is important for military applications, and it is also used in applications in the regulated sector such as aviation, railroad systems, and medical devices. Ada 83 included the package construct, which supports encapsulation and modularization. Ada 95 provided comprehensive support for object-oriented programming, through classes, polymorphism, inheritance, and dynamic binding. Ada supports concurrent programming, and the unit of concurrency is a program entity known as a task, and tasks can communicate either asynchronously or synchronously.

Chapter 4
Agile Methodology

Agile is a recent innovation in the software engineering field, and it is a popular lightweight software development methodology that builds quality into the software. There has been a growth in interest in lightweight software development methodologies since the 1990s, and these include approaches such as rapid application development (RAD), dynamic systems development method (DSDM), and extreme programming (XP). Agile provides opportunities to assess the direction of the project throughout the software development lifecycle, and it has a strong collaborative style of working..

The *Agile* methodology is more responsive to customer needs than traditional methods such as the waterfall model. *The waterfall development model is like a wide and slow-moving value stream* and halfway through the project 100% if the requirements are typically 50% done. *However, for Agile development 50% of requirements are typically 100% done halfway through the project.*

An early version of the Agile methodology was introduced in the late 1980s/ early 1990s, and the methodology has a strong collaborative style of working, with every aspect of development such as requirements and design continuously revisited during the development, and the direction of the project is regularly evaluated.

Agile focuses on rapid and frequent delivery of partial solutions developed in an iterative and incremental manner. Each partial solution is evaluated by the product owner, and the feedback is used to determine the next steps for the project. Agile is more responsive to customer needs than traditional methods such as the waterfall model, and its adherents believe that it results in:

- Higher quality.
- Higher productivity.
- Faster time to market.
- Improved customer satisfaction.

It advocates adaptive planning, evolutionary development, early development, continuous improvement, and a rapid response to change. The term *Agile* was coined by Kent Beck and others in the Agile Manifesto in 2001 (www.agilealliance.

© Springer Nature Switzerland AG 2018

G. O'Regan, *The Innovation in Computing Companion*,
https://doi.org/10.1007/978-3-030-02619-6_4

org) (Beck et al. 2001). The traditional waterfall model is like a wide and slow-moving value stream, and it is difficult and time-consuming to correct defects and missing requirements that are identified in the later part of the development lifecycle.

Agile has a strong collaborative style of working, and ongoing changes to requirements are considered normal in the Agile world. It argues that it is more realistic to change requirements regularly throughout the project rather than attempting to define all requirements at the start of the project (as in the waterfall methodology). Agile includes controls to manage changes to the requirements and good communication with the stakeholders, and regular feedback is an essential part of the process.

A *user story* may be a new feature or a modification to an existing feature. The feature is reduced to the minimum scope that can deliver business value, and a feature may give rise to several stories. Stories often build upon other stories, and the entire software development lifecycle is employed for the implementation of each story. Stories are either done or not done (i.e., there is no such thing as 50% done), and the story is complete only when it passes its acceptance tests.

Scrum is an Agile method for project managing iterative development, and it consists of an outline planning phase for the project, followed by a set of sprint cycles (where each cycle develops an increment). *Sprint planning* is performed before the start of the iteration, and stories are assigned to the iteration to fill the available time. *Each scrum sprint is of a fixed length (usually 2–4 weeks), and it develops an increment of the system.*

The estimates for each story and their priority are determined, and the prioritized stories are assigned to the iteration. A short (usually 15 minutes) morning *stand-up meeting* is held daily during the iteration, and it is attended by the scrum master, the project manager,[1] and the project team. It discusses the progress made the previous day, problem reporting and tracking, and the work planned for the day ahead. A separate meeting is held for issues that require more detailed discussion.

Once the iteration is completed the latest product increment is demonstrated to a review audience including the product owner. This is to receive feedback and to identify new requirements. The team also conducts a retrospective meeting to identify what went well and what went poorly during the iteration, as part of continuous improvement for future iterations.

The planning for the next sprint then commences. The *scrum master* is a facilitator who arranges the daily meetings and ensures that the scrum process is followed. The role involves removing roadblocks so that the team can achieve their goals and communicating with other stakeholders. Agile employs *pair programming* and a collaborative style of working with the philosophy that two heads are better than one. This allows multiple perspectives in decision-making which provides a broader understanding of the issues.

---

[1] Agile teams are self-organizing, and small teams (team size <20 people) generally do not usually have a project manager role. The scrum master performs some light project management tasks.

Software testing is important in verifying that the software is fit for purpose, and Agile generally employs automated testing for unit, acceptance, performance, and integration testing. Agile employs *test-driven development* with tests written before the code, and the developers write code to make a test pass with ideally developers only coding against failing tests. This approach forces the developer to write testable code, as well as ensuring that the requirements are testable. Tests are run frequently with the goal of catching programming errors early. They are generally run on a separate build server to ensure that all the dependencies are checked. Tests are rerun before making a release.

Refactoring is a design and coding practice employed in Agile. Its objective is to change how the software is written without changing what it does. Refactoring is a tool for evolutionary design where the design is regularly evaluated, and improvements are implemented as they are identified. *Refactoring helps in improving the maintainability and readability of the code and in reducing complexity*. The automated test suite is essential in demonstrating that the integrity of the software is maintained following refactoring.

Continuous integration allows the system to be built with every change. Early and regular integration allows early feedback to be provided, and it also allows all automated tests to be run thereby identifying problems earlier. The Agile philosophy and approach includes:

- Working software is more useful than documents.
- Direct interaction is preferred over documentation.
- Change is accepted as a normal part of life in the Agile world.
- Customer is involved throughout the project.
- Demonstrate value early.
- Feedback and adaptation are employed in decision-making.
- Aim is to achieve a narrow fast-flowing value stream.
- User stories and sprints are employed.
- A project is divided into fixed length iterations (i.e., time boxing is employed).
- Entire software development lifecycle is employed for implementation of a story.
- Stories are either done or are not done (no such thing as 50% done).
- Iterative and incremental development is employed.
- Emphasis on quality.
- Stand-up meetings held daily.
- Rapid conversion of requirements into working functionality.
- Delivery is made as early as possible.
- Maintenance is considered to be part of the development process.
- Refactoring and evolutionary design employed.
- Continuous integration is employed.
- Short cycle times.
- Plan regularly.
- Early decision-making.

Stories are prioritized based on several factors including:

- Business value of story.
- Mitigation of risk.
- Dependencies on other stories.

The key parts of Agile are described in Table 4.1 below.

Quality management in Agile differs from traditional quality management of software projects, which is focused on verifying compliance to the defined process and the project quality plan. The assumption in traditional quality management is that if the process is faithfully followed, then product quality will follow. However, traditional quality management just guarantees that the process is followed and does not guarantee that quality has been achieved. Further, even if the product fully complies with the specification, it may fail to be what the customer wants.

Agile is focused on building quality into the software, and it works well when the right product owner is involved, sufficient time is allowed in developing the software, and good collaboration techniques are employed in the process. It is important to ensure that the right people are involved in the process and that the product is fit for purpose. It is important to determine that the Agile project achieves the criteria to deliver the "built-in" quality, although it is a little unclear how to assess project quality in Agile and as to what constitutes a quality Agile delivery.

Of course, as in traditional project delivery, Agile projects can be subject to time pressures that may potentially compromise quality, where there may be conflicts between getting the product complete quickly versus getting it done right. Agile projects need to resist these pressures to ensure that quality is delivered and ensure that the good Agile practices are fully carried out.

Table 4.1 Features of Agile

Area	Description
Scrum	*Scrum* is a framework for managing an agile software development project. It breaks the software development into a series of sprints (of 2–4 weeks). The team creates a sprint backlog (to do list); they attend a daily 15-min stand-up meeting; the main deliverables are the product itself, product backlog, sprint backlog, sprint burnout, and release burnout charts. The scrum master is the expert on the agile process and acts as a coach to the team
User story	A *user story* is a short simple description of a feature written from the viewpoint of the user of the system. They are often written on index cards or sticky notes and arranged on walls or tables to facilitate discussion. A large story is often split into several smaller user stories
Estimation	Planning poker is a popular consensus-based estimation technique often used in agile, and it is used to estimate the effort required to implement a user story
Test-driven development	The test-driven development of a new feature begins with writing a suite of test cases based on the requirements for the feature, and the code for the feature is then written to pass the test cases
Pair programming	*Pair programming* is an agile technique where two programmers work together at one computer. The author of the code is termed the *driver*, and the other programmer (the *observer*) is responsible for reviewing each line of code

Chapter 5
Amdahl 470 and 580 Computers

Gene Amdahl was the chief architect for the highly successful IBM System/360 (see Chap. 51), and he was appointed an IBM fellow in 1965 in recognition of his contributions to IBM. He became director of IBM's Advanced Computing Systems (ACS) Laboratory in California and given freedom to pursue his own research projects. He later left IBM following disagreements on future computer development, and he formed Amdahl Corporation, which later became a major competitor to IBM in the mainframe market.

His goal was to develop a mainframe that would be compatible with the IBM System/360, and he intended that it would provide superior performance at a lower price. Customers would be able to run IBM applications on Amdahl hardware without buying IBM hardware. He revised his plans following IBM's introduction of its IBM System/370 mainframe, and his new goal was to launch an IBM compatible S/370 mainframe.

Amdahl Corporation initially had difficulty in raising sufficient capital to develop its business. It received funding from Fujitsu who were interested in a joint development program and from Nixdorf who were interested in representing Amdahl in Europe.

Amdahl Corporation launched its first product, the Amdahl 470 V/6, in 1975. This was an IBM System/370 compatible mainframe that could run IBM software, and so it was an alternative to a full IBM proprietary solution. It meant that companies around the world now had the choice of continuing to run their software on IBM machines or purchasing the cheaper and more powerful IBM compatibles produced by Amdahl.

Amdahl's first customer was the NASA Goddard Institute for Space Studies, which was based in New York. The Institute needed a powerful computer to track data from its Nimbus weather satellite, and it had a choice between a well-established company such as IBM and an unknown company such as Amdahl. It seemed likely that IBM would be the chosen supplier. However, the Institute was highly impressed with the performance of the Amdahl 470 V/6, and its cost was significantly less than the IBM machine.

© Springer Nature Switzerland AG 2018
G. O'Regan, *The Innovation in Computing Companion*,
https://doi.org/10.1007/978-3-030-02619-6_5

The Amdahl 470 competed directly against the IBM System 370 family of main-frames. It was compatible with IBM hardware and software but cheaper and more powerful than the IBM product: i.e., the Amdahl machines provided better perfor-mance for less money. Further, the machine was much smaller than the IBM machine due to the use of large-scale integration (LSI) with many integrated circuits on each chip. This meant that the Amdahl 470 was one-third of the size of IBM's 370. It was over twice as fast and sold for about 10% less than the IBM 370.

IBM's machines were water-cooled, while Amdahl's were air-cooled, which decreased installation costs significantly. Machine sales were slow initially due to concerns over Amdahl Corporation's long-term survival and the risks of dealing with a new player. IBM had a long-established reputation as the leader in the com-puter field. The University of Michigan was Amdahl's second customer, and it used the 470 in its education center. Texas A&M was Amdahl's third customer, and they used the 470 for educational and administrative purposes. Amdahl Corporation was well on its way to success, and by 1977 it had over 50 of the 470 V/6 machines installed at various customer sites.

IBM launched a new product, the IBM 3033, in 1977 to compete with the Amdahl 470. However, Amdahl Corporation responded with a new machine, the 470 V/7, which was one and a half times faster than the 3033 and only slightly more expen-sive. Customers voted with their feet and chose Amdahl as their supplier, and by late 1978 Amdahl had sold over 100 of the 470 V/7 machines.

IBM introduced a medium-sized computer, the 4300 series, in early 1979, and in late 1980 it announced plans for the 3081 processor which would have twice the performance of the existing 3033 on its completion in late 1981. In response, Amdahl announced the 580 series (Fig. 5.1), which would have twice the perfor-mance of the existing 470 series. The 580 series was released in mid-1982, but there were some reliability problems with the early processors, and they lacked some of the features of the new IBM product.

Amdahl moved into large system multi-processor design from the mid-1980s. It introduced its 5890 model in late 1985, and its superior performance allowed Amdahl to gain market share and increase its sales to approximately $1 billion in 1986. It now had over 1300 customers in around 20 countries around the world. It launched a new product line, the 5990 processor, in 1988, and this processor outper-formed IBM by 50%. Customers voted with their feet and chose Amdahl as their supplier.

It was clear that Amdahl was now a major threat to IBM in the high-end main-frame market. Amdahl had a 24% market share and annual revenues of $2 billion at the end of 1988. This led to a price war with IBM, with the latter offering discounts to its customers to protect its market share. Amdahl responded with its own dis-counts, and this led to a reduction in profitability for the company.

The IBM personal computer was introduced in the early 1980s, and by the early 1990s, it was clear that the major threat to Amdahl was the declining mainframe market and not IBM. Revenue and profitability fell, and Amdahl shut factory lines and cut staff numbers. By the late 1990s, Amdahl was making major losses, and there were concerns about the future viability of the company.

Fig. 5.1 Amdahl 5860. (Courtesy of Robert Broughton, University of Newcastle)

It was clear by 2001 that Amdahl could no longer effectively compete against IBM following IBM's introduction of its 64-bit zSeries architecture. Amdahl had invested a significant amount in research on a 64-bit architecture to compete against the zSeries, but the company estimated that it would take a further $1 billion and 2 more years to create an IBM-compatible 64-bit system. Further, it would be several years before they would gain any benefit from this investment as there were declining sales in the mainframe market due to the popularity of personal computers.

By late 2001 the sales of mainframes accounted for just 10% of Amdahl's revenue, with the company gaining significant revenue from the sale of Sun servers. Amdahl became a wholly owned subsidiary of Fujitsu in 1997, and it exited the mainframe business in 2002. Today it focuses on the server and storage side as well as on services and consulting. For more detailed information on Gene Amdahl, Amdahl Corporation, and IBM, see (O'Regan 2015).

Chapter 6
Analytic and Difference Engines

The Analytic and Difference Engines were designed by Charles Babbage in the nineteenth century. The Difference Engine was designed to produce mathematical tables for various mathematical functions, where the machine was designed to compute polynomial functions (and many mathematical functions may be approximated by polynomials). The Analytic Engine was designed to execute all tasks expressed in algebraic notation, and it provided the vision of a modern computer. Babbage is considered (along with George Boole) to be one of the grandfathers of computing, and he made contributions to mathematics, statistics, and astronomy.

Babbage (Fig. 6.1) was interested in accurate mathematical tables as these are essential for navigation and scientific work. However, there was a high error rate in the existing tables due to human error introduced during calculation. He was interested in the problem of finding a mechanical method to perform the calculations, with the goal of eliminating errors introduced by human calculation. Pascal invented the *Pascaline* (a simple calculating machine) in the seventeenth century, which was used for performing addition and subtraction. Leibniz later invented a machine called the *Step Reckoner* that could perform addition, subtraction, multiplication, and division. Babbage wished to develop a machine that could compute polynomial functions accurately.

He designed the Difference Engine (No. 1) in 1821 to produce mathematical tables. This was essentially a mechanical calculator (analogous to modern electronic calculators), and it was designed to compute polynomial functions. It could also compute tables for logarithmic and trigonometric functions such as sine or cosine (as these may be approximated by polynomials)[1].

---

[1] The power series expansion of the Sine function is given by $\mathrm{Sin}(x) = x - x^3/3! + x^5/5! - x^7/7! + \ldots$. The power series expansion for the Cosine function is given by $\mathrm{Cos}(x) = 1 - x^2/2! + x^4/4! - x^6/6! + \ldots$ Functions may be approximated by interpolation, and the approximation of a function by a polynomial of degree n requires $n + 1$ points on the curve for the interpolation. That is, the curve formed by the polynomial of degree n that passes through the $n + 1$ points of the function is an approximation of the function. The error function also needs to be considered.

© Springer Nature Switzerland AG 2018
G. O'Regan, *The Innovation in Computing Companion*,
https://doi.org/10.1007/978-3-030-02619-6_6

Fig. 6.1 Charles Babbage

The accurate approximation of trigonometric, exponential, and logarithmic functions by polynomials depends on the degree of the polynomial, the number of decimal digits that it is being approximated to, and on the error function. A higher-degree polynomial is generally able to approximate the function more accurately.

Babbage produced prototypes for parts of the Difference Engine, but he never actually completed the machine. He also designed the Analytic Engine (the world's first mechanical computer). Its design included a processor, a memory, and a way to input information and output results.

6.1   Difference Engine

The difference engine consists of N columns (numbered 1 to N). Each column can store one decimal number, and the decimal digits are represented by wheels. The Difference Engine (No. 1) has seven columns with each column containing 20 wheels. Each wheel consists of 10 teeth which represent a decimal digit, and each column may therefore represent a decimal number with up to 20 digits. The seven columns in Difference Engine (No. 1) allow polynomials of degree six to be represented.

The only operation that the Difference Engine can perform is the addition of the value of column $n + 1$ to column n, and this results in a new value for column n. Column N can only store a constant, and column 1 displays the value of the calculation for the current iteration. The machine is programmed prior to execution by setting initial values to each of the columns. Column 1 is set to the value of the polynomial at the start of computation; column 2 is set to a value derived from the first and higher derivatives of the polynomial for the same value of x. Each of the columns from 3 to N is set to a value derived from the $n-1st$ and higher derivatives of the polynomial.

The first working difference engine was built by George Schuetz and his son Edward in 1853. Their machine was based on Babbage's design, and they produced a prototype in 1843. They later received funding from the Swedish government to complete a working machine, and it could compute polynomials of degree 4 on 15-digit numbers. It was displayed at the World Fair held in Paris in 1855, and George and Edward Scheutz received the gold medal for their contributions. The machine was later sold for £1000 to the Dudley Observatory in the United States. The Science Museum in London has a copy of the third Scheutz Difference Engine on display.

It was the first machine to compute and print mathematical tables mechanically. The machine was accurate, and it showed the potential of mechanical machines as a tool for scientists and engineers.

The Scheutz difference engine was comprised of shafts and wheels. The scientist could set numbers on the wheels[2] and turn a crank to start the computation. The decimal numbering system was employed, and there was also a carry mechanism. The scientist could determine the result of the calculation by reading down each shaft. The difference engine could print out the answers to the computation, and it was the first printing calculating machine.

The machine is unable to perform multiplication or division directly. Once the initial value of the polynomial and its derivatives are calculated for some value of x, the difference engine can calculate any number of nearby values using the numerical method of finite differences. This method replaces computational intensive tasks involving multiplication or division by an equivalent computation that just involves addition or subtraction.

The initial problem is to compute the first row, which allows the other rows to be computed. Its computation is more difficult for complex polynomials. *The second problem is to find a suitable polynomial to represent the function, and this may be done by interpolation.*

However, once these problems are solved, the engine produces pages and columns full of data. Babbage received £17 K of taxpayer funds to build the Difference Engine. However, he only produced prototypes of the intended machine, which were limited to the computation of quadratic polynomials of six digit numbers. Babbage intended that the machine would operate on 6th-degree polynomials of 20 digits. The British government cancelled the project in 1842.

He designed an improved difference engine (No. 2) in 1849 (Fig. 6.2). It could operate on 7th order differences (i.e., polynomials of order 7) and 31-digit numbers. The machine consisted of 8 columns with each column consisting of 31 wheels. It was built over 150 years later (in 1991) to mark the two hundredth anniversary of

---

[2] Each wheel has 10 teeth and represents a decimal digit.

Fig. 6.2 Difference engine No. 2. (Photo public domain)

his birth. The Science Museum in London also built the printer that Babbage designed, and both the machine and the printer worked correctly according to Babbage's design (after a little debugging).

6.2   Analytic Engine

The Difference Engine required human intervention to perform the calculation, and Babbage proposed a revolutionary solution. His plan was to construct a new machine that would execute all tasks that may be expressed in algebraic notation. His Analytic Engine consisted of two parts (Table 6.1).

Babbage intended that the operation of the Analytic Engine would be analogous to the operation of the *Jacquard loom*. The latter is capable of weaving (i.e., executing on the loom) a design pattern that has been prepared by a team of skilled artists. The design pattern is represented by punching holes on a set of cards, where each

Table 6.1   Analytic engine

Part	Function
Store	This contains the variables to be operated upon as well as all those quantities that have arisen from the result of intermediate operations
Mill	The mill is essentially the processor of the machine into which the quantities about to be operated upon are brought

card represents a row in the design. The cards are then ordered and placed in the loom, and the exact pattern is produced by the loom.

The Jacquard loom was the first machine to use punch cards to control a sequence of operations. It did not perform computation, but it was able to change the pattern of what was being weaved by changing cards. This gave Babbage the idea to use punched cards to store programs to perform the analysis and computation in the Analytic Engine.

The use of the punched cards in the Analytic Engine allowed the formulae to be manipulated in a manner dictated by the programmer. The cards commanded the analytic engine to perform various operations and to return a result. Babbage distinguished between two types of punched cards:

– *Operation Cards.*
– *Variable Cards.*

Operation cards are used to define the operations to be performed, whereas the variable cards define the variables or data that the operations are performed upon. The use of punched cards to store programs in the Analytic Engine is similar to the idea of a stored computer program in Von Neumann architecture. However, Babbage's idea of using punched cards to represent machine instructions and data was over 100 years before digital computers. The *Analytic Engine is therefore an important milestone in the history of computing.*

Babbage intended that the program be stored on read-only memory using punch cards and that the input and output would be carried out using punch cards. He intended that the machine would be able to store numbers and intermediate results in memory that could then be processed. There would be several punch card readers in the machine for programs and data. He envisioned that the machine would be able to perform conditional jumps, as well as parallel processing where several calculations could be performed at once.

The Analytic Engine was designed in 1834 as the world's first mechanical computer (Lovelace 1842). It included a processor, a memory, and a way to input information and output results. However, the machine was never built, as Babbage was unable to receive funding from the British Government.

6.2.1 Applications of Analytic Engine

Lady Ada Lovelace (discussed in Chap. 3) produced an annotated translation of Menabrea's *Notions sur la machine analytique de Charles Babbage* (Ada, Augusta, Countess of Lovelace 1842). The notes were about three times the length of the original memoir and considered many of the difficult and abstract questions connected with the subject. They are regarded as a description of a computer and software, and they were published in Richard Taylor's Scientific Memoirs (Vol. 3) in 1843.

She explained in the notes how the Analytic Engine could be programmed, and she wrote what is considered to be the first computer program. This program detailed a plan be written for how the engine would calculate *Bernoulli numbers*. Lady Ada Lovelace is therefore considered to be the *first computer programmer*.

She saw the potential of the Analytic Engine to fields other than mathematics. She predicted that the machine could be used to compose music, produce graphics, as well as solving mathematical and scientific problems. She speculated that the machine might act on other things apart from numbers and be able to manipulate symbols according to rules. In this way, a number could represent an entity other than a quantity, and she believed the Analytic Engine had applications outside of mathematics.

Chapter 7
Apple II and Macintosh Computers

Steve Jobs and Steve Wozniak founded Apple Computers in 1976, with Wozniak responsible for product development and Jobs responsible for marketing. The Apple I computer was released in 1977, and it was mainly of interest to computer hobbyists and engineers. It retailed for $666.66 and generated over $700,000 in revenue for the company. The Apple II computer was released later that year, and it was one of the earliest computers to come preassembled. It included color graphics and came in its own plastic casing (Fig. 7.1). It retailed for $1299 and generated over $139 million in revenue for the company.

The Apple I computer had 4 K of RAM (expandable to 8 K) and 256 bytes of ROM, and it was mainly of interest to computer hobbyists. The user needed to provide a case, power supply, keyboard, and display, as these were not supplied.

The Apple II computer was a significant advance on the Apple I machine, and it was one of the earliest computers to have a color display. It was a popular 8-bit home computer, and the BASIC programming language was built in. It contained 4 K of RAM (which was could be expanded to 48 K). The VisiCalc spreadsheet program was released on the Apple II, which helped to transform it into a credible business machine. The Apple II was a major commercial success, and Apple became a public listed company in 1980. John Scully became CEO of Apple in 1983.

The Apple Macintosh (Fig. 7.2) was announced during a famous television commercial aired during the third quarter of the Super Bowl in 1984. It was one of the most creative advertisements of all time, and it ran just once on television. It generated more excitement than any other advertisement up to then, and it immediately positioned Apple as a creative and innovative company while implying that its competition (i.e., IBM) was stale and robotic.

It presented Orwell's totalitarian world of 1984, with a lady runner wearing orange shorts and a white tee shirt with a picture of the Apple Macintosh running toward a big screen and hurling a hammer at the big brother character on the screen. The audience is stunned at the broken screen, and the voice-over states: *On January 24th Apple will introduce the Apple Macintosh and you will see why 1984 will not*

© Springer Nature Switzerland AG 2018
G. O'Regan, *The Innovation in Computing Companion*,
https://doi.org/10.1007/978-3-030-02619-6_7

Fig. 7.1 Apple II computer. (Photo public domain)

Fig. 7.2 Apple macintosh computer. (Photo public domain)

be like "1984". Ridley Scott directed the short film, and he has also directed well-known films such as Alien, Blade Runner, Robin Hood, and Gladiator.

The Macintosh included a friendly and intuitive graphical user interface (GUI), and the machine was much easier to use than the standard IBM PC. The latter was a command-driven operating system that required the user to be familiar with its operating system commands to carry out the desired tasks.

The Macintosh project began in 1979 with the goal of creating an easy-to-use low-cost computer for the average consumer. It was influenced by the design of the Apple Lisa, and it employed the Motorola 68,000 processor. Steve Jobs became involved in the project in 1981, and he negotiated a deal with Xerox that allowed him and other Apple employees to visit the Xerox PARC research center at Palo Alto in California to see their pioneering work on the Xerox Alto computer. The Alto employed a mouse-driven graphical user interface, and Jobs immediately recognized that the future of computing was in user-friendly GUIs. PARC's research work had a major influence on the design and development of the Macintosh, as Jobs was convinced that all future computers would use a graphical user interface.

The Macintosh was a much easier machine to use than the existing IBM PC, and its friendly and intuitive graphical user interface was a revolutionary change from the command-driven operating system of the IBM PC. The users were not required to remember any operating system commands and could instead navigate around the graphical icons using a mouse. The introduction of the Mac GUI was a major milestone in the computing field, and it was 1990 before Microsoft introduced its Windows 3.0 GUI-driven operating system. The success of the Mac GUI showed the importance of human-computer interaction (HCI) and demonstrated that the usability of a computer needs to be considered as part of its design.

Apple intended that the Macintosh would be an inexpensive and user-friendly personal computer that would rival the IBM PC and compatibles. However, it retailed for $2495, which was significantly more expensive than the IBM PC. Further, initially, it had a limited number of applications available, whereas the IBM PC had spreadsheets, word processors, and database applications. The technically superior Apple Macintosh was unable to break the IBM dominance of the market. However, the machine became very popular in the desktop publishing market, due to its advanced graphics capabilities.

The sales of the Macintosh were slow, and Apple went through financial difficulty in the mid-1980s. This led to boardroom tensions at senior management level and a power struggle between Jobs and Scully. Jobs resigned from Apple in 1985 and founded NeXT, Inc. in 1986 and Pixar[1] (a web animation company) in 1987. NeXT produced the NeXT workstation computers and developed a framework for web application development called WebObjects in the mid-1990s. The Macintosh OS X and *i*Phone Operating System (iOS) were later based on the NeXTSTEP operating system. NeXT was acquired by Apple for over $400 million in 1997, and this led to Jobs return to Apple, and he became CEO later that year with an annual salary of $1.[2]

The *i*Mac (a Macintosh desktop computer) was released in 1998, where the letter *i* stands for the *Internet*, and it also represents the fact that the product is a personal device designed for the *individual*. The *i*Mac originally employed the Power PC

---

[1] Pixar was acquired by Disney in 2006 for $7.4 billion, with Jobs (who owned 50.1% of Pixar) becoming Disney's largest shareholder (with 7% of the company and a seat on the board).

[2] Jobs joked that $0.50 was for salary and $0.50 was for performance. He was also given generous share options.

chip designed and developed by IBM and Motorola, but these were later replaced with Intel processors in 2006. The entire Macintosh line was transitioned to Intel processors in 2006.

John Sculley, the CEO of Apple, coined the term *Personal Digital Assistant*, and Apple introduced the first PDA, the Newton, in 1993. The Apple Newton had some nice features such as limited handwriting recognition abilities. Xerox PARC had created a prototype PDA, the Dynabook, in the 1970s, but they did not commercialize it. A PDA allows a large amount of data to be stored on a small handheld device.

The *i*Book (a line of personal laptops) was introduced in 1999, and the *i*Pod (a portable music player) was introduced in 2001. The *i*Pod is a small portable hard disk MP3 player which has a capacity of 5–10 GB, and it can hold up to 1000 MP3 songs. The *i*Pod prepared the way for the digital media Apple *i*Tunes Store, which is the largest music vendor in the world (with over 40 million songs available).

The *i*Tunes Music Store was launched in 2003; it allows songs to be downloaded for a small fee. The individual songs are sold for the same price and without a subscription fee for access to the catalogue. There are 35–40 million songs in the catalogue, and movies and TV shows are also available for purchase. The store has sold billions of songs worldwide.

Apple entered the mobile phone market with the release of the *i*Phone in 2007. This Internet-based multimedia smartphone included a touch screen and features such as a video camera, email, web browsing, text messaging, and voice. The *i*Phone had a 3.5 inch 480×320 touch screen, a QWERTY keyboard, and a 4GB of storage. Apple developed its own operating system, *i*OS, for the *i*Phone, and this revolutionary Internet-based smartphone with its touch screen became an immediate success.

Apple's *i*Phone 4 smartphone (Fig. 7.3) was introduced in 2010, and it has a 3.5 inch 960×640 screen and a 5-mega pixel camera. Apple released the *i*Pad in 2010, which is a large screen tablet-like device that uses a touch screen operating system. Jobs suffered from ill health from 2003 when he was diagnosed with a rare type of pancreatic cancer. He fought the disease for several years and died in 2011. For more information on Steve Jobs and Apple see (Isaacson 2011).

Fig. 7.3 Apple *i*Phone 4.
(Photo public domain)

Chapter 8
Artificial Intelligence and Applications

The long-term goal (perhaps hundreds of years) of artificial intelligence is to create a thinking machine that is intelligent, has consciousness, has the ability to learn, has free will, and is ethical. The field involves several disciplines such as philosophy, psychology, linguistics, machine vision, cognitive science, logic, and ethics. It is a young field, and the term was coined by the organizers of the 1956 Dartmouth workshop on artificial intelligence. Some researchers (e.g., Hubert Dreyfus and John Searle) believe that its goals are impossible or incoherent, and others (e.g., Joseph Weizenbaum) argue that the field is unethical.

The success of early AI went to its practitioners' heads, and they believed that they would soon develop machines that would emulate human intelligence. They convinced many of the funding agencies and the military to provide research grants, as they believed that real artificial intelligence would soon be achieved. They had some initial (limited) success with machine translation, pattern recognition, and automated reasoning. However, it soon became clear that AI is a long-term multidisciplinary project.

Alan Turing devised the famous *Turing Test* to judge whether a machine is conscious and intelligent (Turing 1950). Turing's 1950 paper was very influential as it raised the idea of the possibility of programming a computer to behave intelligently. Searle's Chinese Room thought experiment is a famous rebuttal of machine understanding, and it rejects the claim that a machine will someday in the future have the same cognitive qualities as humans.

8.1 The Turing Test

The Turing Test is an adaptation of a party game, which involves three participants. One of them, the judge, is placed in a separate room from the other two: one is male and the other is female. Questions and responses are typed and passed under the door. The objective of the game is for the judge to determine which participant is

© Springer Nature Switzerland AG 2018
G. O'Regan, *The Innovation in Computing Companion*,
https://doi.org/10.1007/978-3-030-02619-6_8

male and which is female. The male can deceive the judge, whereas the female is supposed to assist.

Turing adapted this game to allow a computer to play the role of the male. The computer is said to pass the Turing test if the judge is unable to determine which of the participants is human and which is machine. Turing's 1950 paper on machine intelligence was controversial (Turing 1950), as defenders of traditional values attacked the idea of machine intelligence. Turing strongly believed that machines would eventually be developed that would stand a good chance of passing the test.

The viewpoint that a machine will one day pass the Turing Test and be considered intelligent is known as *strong AI*. It states that a computer with the right program would have the mental properties of humans. There are several objections to strong AI, and one well-known rebuttal is *Searle's Chinese Room* argument (Searle 1980).

8.2   Searle's Chinese Room

Searle's Chinese Room is a famous thought experiment and rebuttal of machine understanding. A man is placed into a closed room into which Chinese writing symbols are input to him (Fig. 8.1). He is given a rulebook that shows him how to manipulate the symbols to produce Chinese output. He has no idea as to what each symbol means, but with the rulebook, he can produce the Chinese output.

This allows him to communicate with the other person and appear to understand Chinese. The rulebook allows him to answer any questions posed, without the slightest understanding of what he is doing or what the symbols mean. The operation of the Chinese room is as follows:

1. Chinese characters are entered through slot 1 (input)
2. The rulebook is employed to construct new Chinese characters
3. Chinese characters are outputted to slot 2 (output)

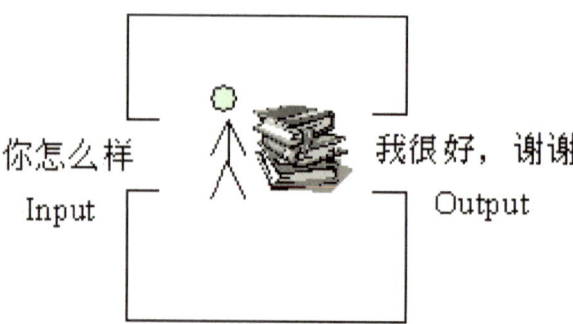

Fig. 8.1 Searle's Chinese room

The question *Do you understand Chinese?* could potentially be asked, and the rulebook would be consulted to produce the answer *Yes, of course* despite of the fact that the person inside the room has no understanding. It will appear to the person outside the room that the person inside is knowledgeable on Chinese, whereas *the person inside is just following rules without any understanding.*

The process where a person inside the room is using a rulebook to transform the input into the output is essentially what a computer program does in that it takes an input, performs computation based on the input, and then finally produces the output. Searle has essentially constructed a machine that appears to be intelligent but can never be mental. Changing the computer program essentially means changing the rulebook, and this does not increase understanding.

The strong artificial intelligence thesis states that given the right program, *any* machine running it would be mental. However, Searle argues that the program for this Chinese room would have no understanding and that therefore *the strong AI thesis must be false.* Searle's argument shows that the program running on a machine that appears to be intelligent has no real understanding of the symbols that it is manipulating. That is, given any rulebook (i.e., program), the person would never understand the meanings of those characters that are being manipulated.

That is, just because the machine acts like it knows what is going on, it only knows what it is programmed to know. This differs from humans in that humans have consciousness and are aware of their environment, whereas a machine is unaware of its being and situation like humans are. Searle's argument is a compelling rebuttal of strong AI that rejects the argument that machines may have intelligence or consciousness. The Chinese room argument applies to any Turing equivalent computer simulation, and while there are several arguments that seek to rebut Searle's position (e.g., the "System Reply" argument), none are entirely convincing.

8.3   Machine Translation

Linguistics is the theoretical and applied study of language, and it includes the study of phonology, morphology, syntax, semantics, and pragmatics. Noam Chomsky is the father of the field, and he defined the Chomsky hierarchy of grammars (O'Regan 2013a), which classifies grammars into several categories with increasing expressive power for each class.

Early work on computational linguistics began with machine translation work in the United States in the 1950s. The goal was to develop an automated procedure to translate Russian language texts directly into English without human intervention. It was naively believed that it was only a matter of time before automated machine translation would be possible.

However, the initial results were not successful, and it was realized that the automated processing of human languages was complex. This led to the birth of a new field called *computational linguistics* to investigate and develop algorithms and software for processing natural languages. This subfield of AI deals with the comprehension and production of natural languages.

The task of translating one language into another involves decoding the meaning of the source text and encoding this meaning in the target language. It requires an understanding of the syntax and semantics of both languages, as well as the cultural aspects of the translation.

Machine translation has improved in recent years with programs such as *Google Translate* providing useful output. However, an automated high-quality translation of unrestricted text remains a long-term project.

8.4 Driverless Cars

A driverless car (autonomous vehicle) is a vehicle that can sense its environment and navigate its way without human intervention. It uses techniques such as AI, GPS, radar, and computer vision to detect its environment, and it has advanced control systems to determine an appropriate navigation path to its destination. Its navigation needs to be sophisticated to enable it to avoid obstacles and to observe road signage and traffic lights during the journey, as well as dealing with diverse weather/light conditions.

The control systems include sensing and navigation systems, and the analysis of the sensory data must be able to distinguish between different vehicles on the road. The control system must make the correct decisions from the analysis of the images, and this is especially important when dealing with unexpected situations.

Driverless cars will need to be encoded with a moral compass to deal with situations where ethical decisions need to be made. For example, suppose a self-driving vehicle is traveling on a road and two children roll off a grassy bank on to the road. Further, there is no time for the vehicle to brake, and the question is what should the vehicle do where if the vehicle swerves to the left to avoid the children it will hit an oncoming motorbike. *Which decision should the car make and how should it make such a decision*? Further, who should be held accountable when incorrect or unethical decisions are made?

Several technology companies such as Google, Apple, Uber, and Amazon are working on driverless cars, and autonomous vehicles may potentially lead to a significant reduction in road accidents and fatalities. They offer greater mobility for people who cannot operate a vehicle, but there are many challenges and safety/security issues to be solved before the public will have sufficient confidence in their use (Table 8.1).

Table 8.1 Challenges with driverless vehicles

Area	Description
Sensing the surroundings	A motorway looks totally different on a clear day than on a foggy day or at dusk. Driverless cars must be able to detect road features in all conditions, and the sensors need to be reliable
Unexpected encounters	Driverless cars struggle with unexpected situations (e.g., traffic police waving vehicles through a red light), as rule-based programming is unlikely to cover every scenario
Human-vehicle interaction	Most self-driving cars will be semiautonomous for the foreseeable future, and determining the responsibilities of human and machine and when one or the other should be in control is a challenge
Ethical	Should the car prioritize the protection of the pedestrian or the passenger? Moral judgments may be required
Security/ hackling	Conventional vehicles have vulnerabilities that may be exploited by hackers (e.g., the braking and steering system of a vehicle was hacked through its entertainment system in 2015). Self-driving cars have more vulnerabilities and are at greater risk to a malicious attack
Legal framework/ liability	Self-driving vehicles will be subject to strict safety regulations, and appropriate legislation needs to be developed

Chapter 9
Atari Video Games

Atari, Inc. is legendary in the world of video games, and the company also designed and manufactured home computers. Atari laid the foundation for the modern video game industry, and it developed video games such as Pong, Asteroids, Tempest, Centipede, and Star Wars.

The company was founded by Nolan Bushnell (Fig. 9.1) and Ted Dabney in 1972. They had founded the engineering firm, Syzgy Engineering in 1971, and the company designed and developed the first arcade video game, *Computer Space*, later that year. This computer game was functionally quite like an early computer game called Spacewar[1], and it was not entirely successful, as it was perceived as being a little complicated to use. However, it still sold over 1500 units, and Syzgy Engineering was incorporated as Atari Inc. in 1972. The name *atari* is used in Japanese when a prediction comes through or when someone wins the lottery. It comes from the Japanese verb *ataru* which means "to hit a target," and it is associated with good fortune.

Bushnell developed a fascination for one of the earliest video games, *Spacewar*, while he was a student at the University of Utah. Steve Russell and others developed this game on a Digital PDP-1 computer at MIT in the early 1960s.

The field of computer graphics emerged with the development of computer graphics hardware, and Ivan Sutherland of MIT (and later the University of Utah) played an important role. He developed sketchpad software in the late 1950s that allowed a user to draw simple shapes on the computer screen, and he invented the first computer-controlled head-mounted display (HMD) in the mid-1960s. The University of Utah became the leading research center in computer graphics in the late 1960s, and so Bushnell received a solid foundation in the computer graphics field.

Bushnell had worked in an amusement arcade during his school holidays, and it occurred to him that a video game could potentially operate as a coin-operated machine. The existing arcades were dominated by coin operated machines such as

---

[1] The Spacewar game was developed by Steve Russell and others at MIT in the early 1960s.

© Springer Nature Switzerland AG 2018
G. O'Regan, *The Innovation in Computing Companion*,
https://doi.org/10.1007/978-3-030-02619-6_9

Fig. 9.1 Nolan Bushnell

pinball cabinets, slot machines, and other trivial games of skill and chance. Bushnell's vision was that of an arcade that would contain coin-operated video games, which would inspire and challenge teenagers. Atari went on to become one of the leading companies in video games.

Atari Inc. hired Al Alcorn as its first design engineer, and he had previously worked with Bushnell and Dabney. Alcom had no background in computer game development, and Computer Space was the first computer game that he had seen. Nevertheless, he designed and developed Pong (Fig. 9.2), which was an arcade version of an existing tennis game[2] for the Magnavox Odyssey home video game console. Bushnell had attended the demonstration of this first ever home video game console, and Alcom made significant improvements to Magnavox's existing game. The new game was called "Pong," and it was a sports game that simulated table tennis. The player could compete against a computer or against another player. Alcom's improvements included speeding up the ball the longer the game went on and adding sound. Pong became popular very quickly, and digital table tennis became highly addictive.

Pong was a commercial success with a single unit earning approximately $40 per day, and Atari was soon receiving orders faster than it could deliver them. Over 8000 machines were delivered to bars, amusement arcades, and other places around the world by 1974. Pong showed that a coin-operated video game could be both

---

[2]Atari later settled a court case brought against it by Magnavox over alleged patent infringement of Magnavox's Odyssey tennis game. Magnavox won millions in various patent disputes, and Atari became a licensee of Magnavox.

Fig. 9.2 Original Atari
Pong video game console

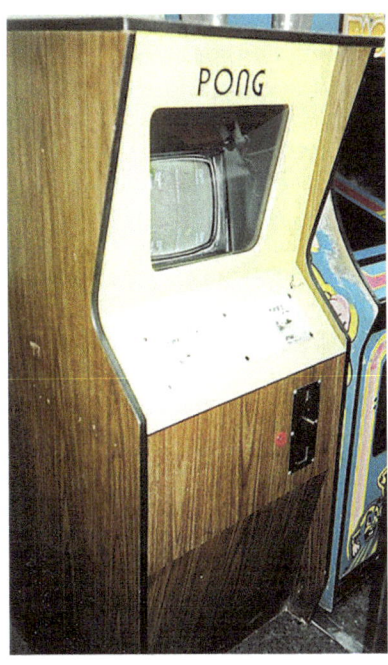

Fig. 9.3 Atari Video
Computer System (VCS)

popular and profitable. The home version of Pong was released in 1975, and it sold 200,000 units in its first year.

Atari did not have any patents protecting Pong, and soon other vendors were offering imitation products. Atari continued to innovate and released successful Atari cabinets including Space Race, Tank, Gotcha, and Breakout in the mid to late 1970s.

Atari had been looking for a way to bring all its existing arcade game to the home market. It designed the Atari Video Computer System (VCS) in 1977, which was later marketed as the Atari 2600 (Fig. 9.3). This was a home game console, which used the MOS 6052 microprocessor, and it provided an affordable way for high-quality video games to be played at home. There were significant financial costs associated with the development and manufacture of the VCS, and Bushnell made a strategic decision to sell Atari to Warner for $26 million to secure the required funding. The Atari VCS was introduced in 1977, and it was priced at $199. It would eventually become one of the most successful video game consoles, but its initial sales were quite low.

Bushnell left the company in 1978 and Ray Kassar took over. By 1979, over a million units of the Atari 2600 were sold, and over 10 million units were sold in 1982. Atari entered the home computer market in 1979 with its release of the Atari 400 and 800 8-bit home computers.

However, there were deep problems at Atari despite the success of the Atari 2600. Warner did not fundamentally understand a technology business, and management alienated many of the creative software development staff. Several of Atari's key engineers left the company to form Activision, a new company that made third-party games for the VCS. Activision's games were better than Atari's, and third-party software developers were also creating games specifically for the Atari 2600. This helped sales of the Atari 2600 to soar, but the quality of the games being produced by Atari began to deteriorate.

Atari now had three areas of business: its arcade business, its home video game business, and its home computer business. However, these three business areas were not working closely together, and Warner did not invest sufficiently in new technology and product development for future success. This was to prove fatal for the company.

The market reaction to Atari's release of Pac-Man and E. T., The Extra Terrestrial, in 1982 was very negative, and Atari was left with a large quantity of unsold inventory that depressed prices. Atari's problems were compounded with the video game crash of 1983, and it lost over $300 million in the second quarter of that year. It was also facing major challenges in the home computing market with users moving from game machines to home computers. Arcades had become less important as video games were now being played at home, and Atari was failing to innovate with new products.

Warner sold the home computing part of the Atari business to Jack Tramiel[3] in 1984, and Tramiel later renamed it to Atari Corporation. Atari Corporation developed and sold video game consoles, video games developed for home use, as well as home and personal computers.

Warner held on to its arcade business until 1985 when it is sold to Namco. Atari's arcade business faded into obscurity, but Atari Corporation continued in business as a designer of home and personal computers until the early 1990s.

---

[3] Jack Tramiel was the founder of Commodore Business Machines.

9.1   Atari Computers

Atari designed and produced four lines of home and personal computers from the late 1970s up to the early 1990s. These were the 8-bit Atari 400 and 800, the 16-bit ST line, the IBM PC compatible series, and the 32-bit series.

The Atari 8-bit series began as a next-generation follow-up to its successful Atari 2600 Video Game Console. Atari's management noted the success of Apple in the early personal computer market, and they tasked their engineers to transform the hardware into a personal computer system. The net result was the Atari 400 and the Atari 800 home computers, which were introduced in 1979. The Atari 800 came with 8 KB of RAM, and the Atari 400 was a lower specification version. These products made an impact on home computing from the early 1980s.

Jack Tramiel acquired Atari's home computing division in 1984, and he renamed the company to Atari Corporation. The company designed the 16-bit GUI-based personal computer, the Atari ST, in 1985, which included a 360 KB floppy disk drive, a mouse, and a monochrome monitor. The Atari ST included two Musical Instrument Digital Interface (MIDI) ports, which made it very popular with musicians.

The Atari 1040 ST was introduced in 1986, and this 16-bit machine contained 1 MB of RAM and came as a complete system with a base unit, a monochrome monitor, and a mouse. The Atari ST line had an impressive life span starting in 1985 and ending with the Atari Mega STE in 1990.

Atari released its first IBM-compatible personal computer, the Atari PC, in 1987. This machine was an 8 MHz 8088 machine with 512 KB of RAM and a 360 KB 5.25-inch floppy disk drive in a metal case. It released the Atari PC2 and PC3 later that year, and the PC3 included an internal hard disk.

The Atari ABC (Atari Business Computer) was released in 1990. The 286 version had several choices of CPU and storage, ranging from an 8 MHz to a 20 MHz CPU and a 30 MB to 60 MB hard disk. The Atari ABC 386 version included a 20 MHz or 40 MHz CPU and a 40 MB or 80 MB hard disk. The ABC 386 shipped with Microsoft Windows 3.0.

Atari shut down its home computer business in 1993 to focus on making its own console. It released the Atari Jaguar, the first 64-bit console, in 1993, and although this was a powerful console for its time, it had a poor controller and poor software support. It failed to succeed against competitor products such as Super Nintendo and Sega Genesis. Atari Corporation ceased trading in 1996, and the reader may consult (Edwards 2011; IGN presents the history of Atari 2014) for more detailed information on the history of Atari and Atari computers.

Chapter 10
Automated Teller Machine

An Automatic Teller Machine (ATM) is an electronic communication device that allows customers of a financial institution to access their account and to perform financial transactions such as checking their account balance and withdrawing cash without the need for a bank teller. The customer is issued with a plastic card with a magnetic stripe (or a chip) that contains card number and security information that uniquely identifies the customer. The customer inserts the card into the machine and enters their personal identification number (PIN), and the PIN entered must match the PIN stored in the card. There are several million ATMs in use around the world (Fig. 10.1).

The first operational ATM was installed at a branch of Barclays Bank in Enfield Town, London, in 1967. John Adrian Shepherd-Barron, who was the managing director of De La Rue Plc, invented the ATM machine. This company manufactures banknotes in over 150 national currencies and specializes in high-quality printing of secure documents such as passports and driving licenses.

Shepherd-Barron believed that there must be a way to get your own money from a bank even if the bank was closed and that this should be possible from anywhere in the United Kingdom and around the world. He was inspired by the fact that a chocolate vending machine dispenses chocolate in any geographical location and that similarly a cash-dispensing machine should be capable of dispensing cash anywhere in the world.

The De La Rue Automated Cash System (DACS) used check-like tokens that contained security information, which was then matched against the PIN code entered by the user on the keyboard. Shepherd-Barron received the Order of the British Empire (OBE) in 2005 for services to banking and for inventing the automatic cash dispenser.

The Scottish inventor, James Goodfellow, was actively involved in the development of an alternate cash-dispensing machine. His machine appeared shortly after Shepherd-Barron's invention, but it is closer to modern ATM machines. Goodfellow patented PIN technology, and his machine accepted a machine readable encrypted card when the customer entered the correct PIN using the numerical keypad.

© Springer Nature Switzerland AG 2018
G. O'Regan, *The Innovation in Computing Companion*,
https://doi.org/10.1007/978-3-030-02619-6_10

Fig. 10.1 Wincor-Nixdorf
Procash 2050 cash
dispenser

Table 10.1 Parts of an ATM

Part	Description
Card reader	The card reader reads the information that is stored on the magnetic strip (or chip) on the card (e.g., account and PIN information) and passes the information to a host processor to retrieve the customer account information
Keypad	This allows customers enter information such as their PIN code, the transaction to be performed, and the amount of the transaction to be carried out
Display screen	This allows the customer to see each step of the process
Speaker	This provides the customer with an audio signal when a key is pressed on the keypad
Printer	This allows a paper receipt of the financial transaction to be printed (should the customer require one)
Cash dispenser	The safe and cash-dispensing mechanism are located deep inside the ATM machine (typically in a safe that is attached to the floor), and cash is dispensed when requested by a customer

The first modern ATM was an IBM 2984 Cash Issuing Terminal developed for Lloyds Bank in England in late 1972. It was the first true ATM and is like the machines in use today. It was named *Cashpoint*, which is a registered trademark of Lloyds. It is often used as the generic trademark for all ATMs used in the United Kingdom.

The components of an ATM consist of a CPU, a magnetic (or chip) card reader to identify the customer, a keypad, function key buttons or a touchscreen to select the financial operation to perform, a display, a printer for printing the receipts, and a vault which is attached to the floor for security purposes to prevent theft (Table 10.1).

Most modern ATMs are connected to the interbank network, which allows customers to make withdrawals from machines that do not belong to the bank where they have their account. This allows withdrawals to take place in different countries

Fig. 10.2 Basic ATM
architecture

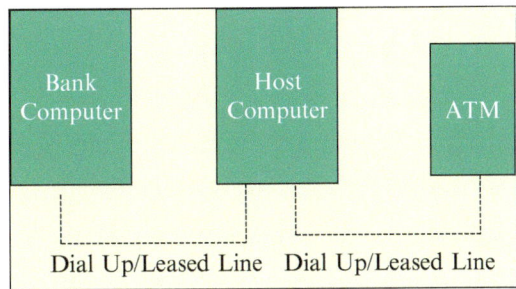

in the local currency of the country. Two popular interbank networks are the PLUS and Cirrus networks, which deal with payments to Visa card and Mastercard holders, respectively. Many banks apply ATM usage fees to noncustomers, and sometimes the fees are applied to all users.

ATMs connect to their ATM controller (ATMC) on the host computer via dialup modems or leased lines, and the ATM controller is used to route financial transactions between the ATM, the core banking systems, and other banks (Fig. 10.2). Many ATMs employ a personal computerlike architecture and run the Microsoft Windows operating system.

The security of the ATM is a key concern, and the goal is to make the ATM invulnerable to hostile attacks. This could include a physical attack where the perpetrators attempt to gain access to the safe where the cash is stored, and so the ATM needs to be physically secure and physically attached to the floor.

There has been a move toward a *cashless society* in the western world, and ATMS may need to redefine themselves to remain relevant to modern consumers. Today, ATMs are beginning to offer other services such as allowing customers to lodge checks and cash in the machine or topping up their mobile phone. It is conceivable that in the future an ATM may have an option to interact directly with banking staff via video banking or to essentially behave like a bank in a box.

Chapter 11
AXE System

L.M. Ericsson is one of the oldest established telecommunication providers, and this Swedish multinational is a leading manufacturer of switching equipment with plants in Europe, Asia, and North America. It is a world leader in communications technology and in the mobile network infrastructure market, and it has a long and distinguished history in fixed network and mobile network communication.

Lars Magnus Ericsson founded the company in Stockholm in 1876, and it initially made telegraph equipment for the state railway. Ericsson later began manufacturing telephones and telecommunication equipment, and it became the leader in fixed line phone technology in the late 1970s. It became the leader in mobile cellular technology from the late 1980s, and it jointly developed (with Televerket) the AXE automatic digital system. It manufactured the AXE digital telephone exchanges from the late 1970s, and AXE has been deployed in many telephone exchanges around the world (Fig. 11.1).

The AXE (Automatic Exchange Electric) switching system was the first fully automated digital switching system, and it converted speech into digital (i.e., the binary language used by computers). It was introduced in 1977, and Ericsson's competitors were still using the slower and less reliable analog systems. AXE is a stored program control telephone switching system designed for a wide number of exchanges.

The older analog system used an electric current to convey the vibrations of the human voice, whereas the AXE digital system uses a stream of binary digits to represent sound. The AXE system was an immediate success with telecom companies, and it is used in many countries around the world. It was originally a digital exchange for landline telephony, but it was later enhanced for use with mobile telephony systems.

Ellemtel was established in 1970 as a pure research and development company, and it was formed as a joint venture between Televerket (Sweden's state-owned PTT) and Ericsson. Its primary task was to develop an electronic automated switching system for telephone stations (and this would become the AXE system).

© Springer Nature Switzerland AG 2018
G. O'Regan, *The Innovation in Computing Companion*,
https://doi.org/10.1007/978-3-030-02619-6_11

Fig. 11.1 AXE system.
(Courtesy of Ericsson)

Ericsson had been working on a commercial electronic switching system called AKE, while Televerket was working on its own electronic switch. Ericsson's proto-type AKE exchange was opened in the town of Tumba in 1968, and an operational exchange was opened in Rotterdam in 1971. It could handle 2400 incoming trunks and 2400 outgoing trunks. A larger exchange was opened in Copenhagen in 1974.

However, the AKE system was not suitable for large switching stations, and Ericsson realized that it needed to develop a new generation of switching systems. It decided to combine its resources with Televerket, and together they formed Ellemtel, which was jointly owned by both companies. Ellemtel's mission was to conduct research and development work on electronic exchanges and digital trans-mission. It was responsible for all research and development of the AXE system, and Ericsson was responsible for manufacturing and marketing.

Bengt-Gunnar Magnusson was the project manager of the AXE project, and AXE had a modular system design which made the system flexible and capable of evolving over time for use in different types of exchanges. New functionality could be added and existing modules updated or replaced. The modular design enabled the system to be easily adapted to different markets. Each module acted as a "black box" and is designed to perform specific function independent of its implementa-tion[1]. The module is essentially defined by its interface, and the software in AXE plays a key role with the hardware providing simple functions.

---

[1] This is the information hiding principle that was introduced by Parnas in the early 1970s.

The development of AXE involved the development of hardware and software such as programs and processors to control the AXE stations. The first prototype AXE system was installed at a Televerket station in Sodertalje near Stockholm in 1976, and it was operational at the exchange in early 1977. Ellemtel's work in developing the AXE system was complete in 1978.

The AXE system was then commercialized and many of Ellemtel's employees moved to Ericsson. AXE was an immediate success, and Ericsson soon had customers in Sweden, Finland, France, Australia, and Saudi Arabia. The Saudi order was the largest that Ericsson had ever received, and it involved increasing the capacity of the Saudi network by 200% and installing the AXE system.

The introduction of AXE meant that by the early 1980s that Ericsson had the market's most advanced and flexible switching system, and it was ideally placed for the transition to mobile telephony. It meant that Ericsson had moved from being a minor player in the telecom business to a major league player. Ericsson was now the leader in fixed line phone technology, and it had laid the foundation for its future success in mobile telephony, where it became the leader in mobile technology from the late 1980s.

The AXE system provided the foundation for Ericsson's growth in mobile telephony. Its flexible modular design allowed new functionality to be added, and by changing a module, AXE was reconfigured to handle mobile telephone calls. This allowed Ericsson to design the first mobile telephone exchange (MTX) by replacing the subsystem for fixed subscribers with a new subsystem for mobile subscribers. The MTX switch was developed in the late 1970s/early 1980s and was a key part of the Nordic Mobile Telephone system (NMT) which would be used in all Nordic countries.

Ericsson was awarded a large Saudi Arabian contract to deliver a fixed line and mobile system, and it was agreed that the NMT standard would be used and that Ericsson would supply the entire system. The Saudi mobile phone network became operational from 1981, and Ericsson provided base stations, radio towers, and switches. Ericsson had now acquired cell-planning experience, and it was awarded the contract to develop the entire mobile telephone network in the Netherlands. It was now a total systems supplier in mobile telephony, providing the entire infrastructure such as switches and base stations. Today, its base stations range from small picocells to large macrocells.

Ericsson began to manufacture mobile phones in the late 1980s, and this was a major change from sales of network infrastructure equipment to the network operator of the country, to sales of mobile phones to individual consumers in each country. It became a leading player in the development of mobile phones, and Nokia, Ericsson, and Motorola dominated the market for mobile phones in the late 1990s. However, while Ericsson remains a leader in network infrastructure (despite massive competition from Huawei and other players), its mobile phone sales have experienced a serious decline. Sony purchased Ericsson's share of the Sony-Ericsson venture in 2012, and Ericsson is focused on its network infrastructure business. For more information on Ericsson see (Meurling et al. 2001).

Chapter 12
Binary Number System

Arithmetic has traditionally been done using the decimal notation[1], and this positional number system involves using the digits 0,1,2, ... 9. Leibniz[2] was one of the earliest people to recognize the potential of the binary number system[3], and this base 2 system uses just two digits, namely, "0" and "1." Leibniz described the binary number system in his paper *Explication de l'Arithmétique Binaire* published in 1703 (Leibniz 1703), and it describes how binary numbers may be added, subtracted, multiplied, and divided.

The use of binary arithmetic allows more complex mathematical operations to be performed by relay circuits, and Boolean logic (described in Chap. 13) is the perfect model for simplifying such circuits and is the foundation underlying digital computing. The number 2 is represented by 10, the number 4 by 100, and so on. The table of values for the first 15 binary numbers is given in Table 12.1.

The binary number system (base 2) is a positional number system, which uses two binary digits 0 and 1. For example, the binary number 1001.01_2 represents $1 \times 2^3 + 0 \times 2^2 + 0 \times 2^1 + 1 \times 2^0 + 0 \times 2^{-1} + 1 \times 2^{-2} = 1 \times 2^3 + 1 \times 2^0 + 1 \times 2^{-2} = 8 + 1 + 0.25 = 9.25$.

The binary system is ideally suited to the digital world of computers, as a binary digit may be implemented by an *on-off switch*. In the digital world, devices that store information or data on permanent storage media such as disks and CDs or

---

[1] Other bases have been employed in ancient civilisations such as the segadecimal (or base 60) system used by the Babylonians. The decimal system was developed by Indian and Arabic mathematicians between 800 and 900 AD, and it was introduced to Europe in the late twelfth/early thirteenth century. It is known as the *Hindu-Arabic system*.

[2] Wilhelm Gottfried Leibniz was a German philosopher, mathematician, and inventor in the field of mechanical calculators. He developed the binary number system used in digital computers and invented the calculus independently of Sir Isaac Newton. He was embroiled in a bitter dispute with Newton toward the end of his life, with respect to who invented the calculus first.

[3] A binary number system was also described in an ancient Indian Sanskrit text (the Chandahsastra) written around the second century BC, by Pingala.

© Springer Nature Switzerland AG 2018
G. O'Regan, *The Innovation in Computing Companion*,
https://doi.org/10.1007/978-3-030-02619-6_12

Table 12.1 Binary number system

Binary	Dec.	Binary	Dec.	Binary	Dec.	Binary	Dec.
0000	0	0100	4	1000	8	1100	12
0001	1	0101	5	1001	9	1101	13
0010	2	0110	6	1010	10	1110	14
0011	3	0111	7	1011	11	1111	15

temporary storage media such as random-access memory (RAM) consist of many memory elements that may be in one of two states (i.e., on or off).

The digit 1 represents that the switch is on, and the digit 0 represents that the switch is off. Claude Shannon showed in his master's thesis (Shannon 1937) that the binary digits (i.e., 0 and 1) can be represented by electrical switches. This allows binary arithmetic and more complex mathematical operations to be performed by relay circuits and provides the foundation of digital computing.

The decimal system (base 10) is familiar to all from everyday use, and there are algorithms to convert numbers from decimal to binary and vice versa. For example, to convert the decimal number 25 to its binary representation we proceed as follows:

$$
\begin{array}{r|rr}
2 & 25 & \\
\hline
  & 12 & 1 \\
  & 6 & 0 \\
  & 3 & 0 \\
  & 1 & 1 \\
  & 0 & 1 \\
\end{array}
$$

The base 2 is written on the left, and the number to be converted to binary is placed in the first column. At each stage in the conversion the number in the first column is divided by 2 to form the quotient and remainder, which are then placed on the next row. For the first step, the quotient when 25 is divided by 2 is 12 and the remainder is 1. The process continues until the quotient is 0, and the binary representation result is then obtained by reading the second column from the bottom up. Thus, we see that the binary representation of 25 is 11001_2.

Similarly, there are algorithms to convert decimal fractions to their binary representation (to a defined number of binary digits as the representation may not terminate). Further, the conversion of a number that contains an integer part and a fractional part involves converting each part separately and then combining both parts.

The octal (base 8) and hexadecimal (base 16) are often used in computing, as the bases 2, 8, and 16 are related bases and easy to convert between. For example, to convert between binary and octal involves grouping the binary bits into groups of three on either side of the point, as each set of 3 bits corresponds to one digit in the octal representation. Similarly, the conversion between binary and hexadecimal involves grouping into sets of four binary digits on either side of the point. The conversion the other way from octal to binary or hexadecimal to binary is equally simple and involves replacing the octal (or hexadecimal) digit with the 3-bit (or 4-bit) binary representation.

Numbers are represented in a digital computer as sequences of bits of fixed length (e.g., 16 bits, 32 bits). There is a difference in the way in which integers and real numbers are represented, with the representation of real numbers being more complicated.

An integer number is represented by a sequence (usually 2 or 4) of bytes where each byte is 8 bits. For example, a 2-byte integer has 16 bits with the first bit used as the sign bit (the sign is 1 for negative numbers and 0 for positive integers), and the remaining 15 bits represent the number. This means that two bytes may be used to represent all integer numbers between $-32,767$ and $32,767$. A positive number is represented by the normal binary representation discussed earlier, whereas a negative number is represented using 2's complement of the original number (i.e., 0 changes to 1 and 1 changes to 0 and the sign bit is 1). All the standard arithmetic operations may then be carried out (using modulo 2 arithmetic).

The standard arithmetic operations may be performed with binary numbers, including subtraction, multiplication, and division. Addition is the simplest operation, where the addition of two single digit binary numbers is given by:

$$0 + 0 = 0$$
$$1 + 0 = 1$$
$$0 + 1 = 1$$
$$1 + 1 = 0 \left(\text{carry} \, 1 \right)$$

The addition of two binary digits may be implemented with two logic gates (an XOR gate for the sum and an AND gate for the carry). The binary addition of $1 + 1$ results in 10: i.e., the sum is 0, and the carry is 1.

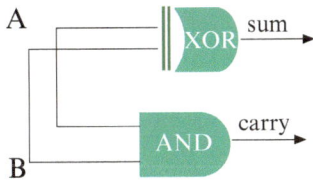

The representation of floating-point real numbers is more difficult, and a real number is represented to a fixed number of significant digits (the significand) and scaled using an exponent in some base (usually 2). That is, the number is represented (approximated as):

$$\text{significand} \times \text{base}^{\text{exponent}}$$

The significand (also called mantissa) and exponent have a sign bit. For example, in simple floating-point representation (4 bytes), the mantissa is generally 24 bits and the exponent 8 bits, whereas for double precision (8 bytes) the mantissa is generally 53 bits and the exponent 11 bits. There is an IEEE standard for floating-point numbers (IEEE 754). Next, we show how binary numbers are used in the computer representation of sets.

12.1   Applications: Computer Representation of Sets

Sets are fundamental building blocks in mathematics, and so the question arises as to how a set is stored and manipulated in a computer. The representation of a set M on a computer requires a change from the normal view that the order of the elements of the set is irrelevant, and we will need to assume a definite order in the underlying universal set M from which the set M is defined.

That is, a set is always defined in a computer program with respect to an underlying universal set, and the elements in the universal set are listed in a definite order. Any set M arising in the program that is defined with respect to this universal set M is a subset of M. Next, we show how the set M is stored internally on the computer.

The set M is represented in a computer as a string of binary digits $b_1 b_2 ..... b_n$ where n is the cardinality of the universal set M. The bits b_i (where i ranges over the values 1, 2, ... n) are determined according to the rule:

$b_i = 1$ if ith element of M is in M
$b_i = 0$ if ith element of M is not in M

For example, if M = $\{1, 2, ... 10\}$, then the representation of $M = \{1, 2, 5, 8\}$ is given by the bit string 1100100100 where this is given by looking at each element of M in turn and writing down 1 if it is in M and 0 otherwise.

Similarly, the bit string 0100101100 represents the set $M = \{2, 5, 7, 8\}$, and this is determined by writing down the corresponding element in M that corresponds to a 1 in the bit string.

Clearly, there is a one-to-one correspondence between the subsets of M and all possible n-bit strings. Further, the set theoretical operations of set union, intersection, and complement can be carried out directly with the bit strings (provided that the sets involved are defined with respect to the same universal set). This involves a bitwise "or" operation for set union, a bitwise "and" operation for set intersection, and a bitwise "not" operation for the set complement operation.

Chapter 13
Boolean Algebra and Digital Computing

George Boole (Fig. 13.1) was born in Lincoln, England, in 1815. His father was a cobbler who was interested in mathematics and optical instruments, and Boole inherited his father's interest in knowledge. He was self-taught in mathematics and Greek, and he taught in various schools near Lincoln. He developed his mathematical knowledge by working his way through Newton's *Principia*, as well as applying himself to the work of mathematicians such as Laplace and Lagrange.

He published regular papers from his early 20s, and these included contributions to probability theory, differential equations, and finite differences. He developed Boolean algebra, which is the foundation for modern computing, and he is considered (along with Babbage) to be one of the grandfathers of computing. *His work was theoretical, and he never actually built a computer or calculating machine. However, Boole's symbolic logic was the perfect mathematical model for switching theory and for the design of digital circuits.*

Boole became interested in formulating a calculus of reasoning, and he published a pamphlet titled *Mathematical Analysis of Logic* in 1847 (Boole 1848). This article developed novel ideas on a logical method, and he argued that logic should be considered as a separate branch of mathematics, rather than a part of philosophy. He argued that there are mathematical laws to express the operation of reasoning in the human mind, and he showed how Aristotle's syllogistic logic could be reduced to a set of algebraic equations. He corresponded regularly on logic with Augustus De Morgan[1], and despite his lack of a formal university qualification, his publications were recognized as excellent[2]. He was awarded the position as the first professor of mathematics at the newly founded Queens College Cork[3] in 1849.

---

[1] De Morgan was a nineteenth-century British mathematician based at University College London. De Morgan's laws state that $(A \cup B)^c = A^c \cap B^c$ and $\neg (A \vee B) = \neg A \wedge \neg B$.

[2] Boole was awarded the Royal Medal from the Royal Society of London in 1844 in recognition of his publications. Sir Rowan Hamilton (inventor of quaternions) was another recipient of this prize.

[3] Queens College Cork is now called University College Cork (UCC) and has about 20,000 students.

© Springer Nature Switzerland AG 2018
G. O'Regan, *The Innovation in Computing Companion*,
https://doi.org/10.1007/978-3-030-02619-6_13

Fig. 13.1 George Boole

His paper on logic introduced two quantities "0" and "1." He used the quantity 1 to represent the universe of thinkable objects (i.e., the universal set), and the quantity 0 represents the absence of any objects (i.e., the empty set). He then employed symbols, such as x, y, z, etc., to represent collections or classes of objects given by the meaning attached to adjectives and nouns. Next, he introduced three operators (+, −, and ×) that combined classes of objects.

The expression xy (i.e., x multiplied by y or $x \times y$) combines the two classes x and y to form the new class xy (i.e., the class whose objects satisfy the two meanings represented by the classes x and y). Similarly, the expression $x + y$ combines the two classes x and y to form the new class $x + y$ (that satisfies either the meaning represented by class x or class y). The expression $x−y$ combines the two classes x and y to form the new class $x−y$. *This represents the class (that satisfies the meaning represented by class x but not class y).* The expression $(1−x)$ represents objects that do not have the attribute that represents class x.

Thus, if x = black and y = sheep, then xy represents the class of black sheep. Similarly, $(1−x)$ would represent the class obtained by the operation of selecting all things in the world except black things; $x\,(1−y)$ represents the class of all things that are black but not sheep; and $(1−x)(1−y)$ would give us all things that are neither sheep nor black.

He showed that these symbols obeyed a rich collection of algebraic laws and could be added, multiplied, etc., in a manner that is similar to real numbers. These symbols may be used to reduce propositions to equations, and algebraic rules may be employed to solve the equations. The rules include:

1. $x + 0 = x$ (Additive identity)
2. $x + (y + z) = (x + y) + z$ (Associative)
3. $x + y = y + x$ (Commutative)
4. $x + (1\text{-}x) = 1$
5. $x1 = x$ (Multiplicative identity)
6. $x0 = 0$
7. $x + 1 = 1$
8. $xy = yx$ (Commutative)
9. $x(yz) = (xy)z$ (Associative)
10. $x(y + z) = xy + xz$ (Distributive)
11. $x(y{-}z) = xy{-}xz$ (Distributive)
12. $x^2 = x$ (Idempotent)

These operations are similar to the modern laws of set theory with the set union operation represented by "+," and the set intersection operation is represented by multiplication. The universal set is represented by "1" and the empty set by "0." The commutative, associative, and distributive laws hold. Finally, the set complement operation is given by $(1{-}x)$.

He applied the symbols to encode Aristotle's syllogistic logic, and he showed how the syllogisms could be reduced to equations. This allowed conclusions to be derived from premises by eliminating the middle term in the syllogism. He refined his ideas on logic further in his book *An Investigation of the Laws of Thought* (Boole 1958). This book aimed to identify the fundamental laws underlying reasoning in the human mind and to give expression to these laws in the symbolic language of a calculus.

He considered the equation $x^2 = x$ to be a fundamental law of thought. It allows the principle of contradiction to be expressed (i.e., for an entity to possess an attribute and at the same time not to possess it):

$$x^2 = x$$

$$\Rightarrow x - x^2 = 0$$

$$\Rightarrow x(1 - x) = 0$$

For example, if x represents the class of horses, then $(1{-}x)$ represents the class of "not-horses." The product of two classes represents a class whose members are common to both classes. Hence, $x(1{-}x)$ represents the class whose members are at once both horses and "not-horses," and the equation $x(1{-}x) = 0$ expresses that fact that there is no such class. That is, it is the empty set.

Boole contributed to other areas in mathematics including differential equations, finite differences, and probability theory. There is an interesting biography of Boole in (McHale 1985). Boole's logic appeared to have no practical use, but this changed with Claude Shannon's 1937 master's thesis, which showed its applicability to switching theory and to the design of digital circuits.

13.1   Switching Circuits and Boolean Algebra

Claude Shannon (Fig. 13.2) showed that Boole's symbolic logic provided the perfect mathematical model for switching theory and for the subsequent design of digital circuits and computers. His influential *master's thesis is a key milestone in computing*, and it shows how to lay out circuits according to Boolean principles. It provides the theoretical foundation of switching circuits, and *his insight of using the properties of electrical switches to do Boolean logic is the basic concept that underlies all electronic digital computers*.

Boolean algebra may be employed to optimize the design of systems of electro-mechanical relays, and circuits with relays solve Boolean algebra problems. The use of the properties of electrical switches to process logic is the basic concept that underlies all modern electronic digital computers. Digital computers use the binary digits 0 and 1, and Boolean logical operations may be implemented by electronic AND, OR, and NOT gates. More complex circuits (e.g., arithmetic) may be designed from these fundamental building blocks.

Shannon showed in his master's thesis *A Symbolic Analysis of Relay and Switching Circuits* (Shannon 1937) that the binary digits (i.e., 0 and 1) can be

Fig. 13.2   Claude Shannon

represented by electrical switches. Switches in circuits may be combined to carry out symbolic logic operations, which allows binary arithmetic and more complex mathematical operations to be performed by relay circuits. He designed a circuit that could add binary numbers, and he later designed circuits that could make comparisons and thus capable of performing a conditional statement. *This was the birth of digital logic and the digital computing age.*

He showed how Boolean algebra could be employed to optimize the design of systems of electromechanical relays and that circuits with relays could solve Boolean algebra problems. His approach provided electronic engineers with the mathematical tool they needed to design digital electronic circuits, thereby providing the foundation of the digital age.

Shannon's master's thesis became the foundation for the practical design of digital circuits, and these circuits are fundamental to the operation of modern computers and telecommunication systems. His insight of using the properties of electrical switches to do Boolean logic is the basic concept that underlies all electronic digital computers.

A circuit may be represented by a set of equations with the terms in the equations representing the various switches and relays in the circuit. Shannon developed a calculus for manipulating the equations using Boole's algebra. The design of a circuit consists of algebraic equations, and these may be manipulated to yield the simplest circuit, which may then be immediately drawn. Complex Boolean value functions can be constructed by combining these digital circuits.

Modern electronic computers use billions of transistors that act as switches and can change state rapidly. A high voltage generally represents the binary value 1 with low voltage representing the binary value 0. A silicon chip may contain billions of tiny electronic switches arranged into logical gates, and these may be combined in various ways to perform more complex tasks such as binary arithmetic. Each gate has binary value inputs and outputs.

The example in Fig. 13.3 is that of an "AND" gate which produces the binary value 1 as output if both inputs are 1, with the result being the binary value 0 otherwise. Figure 13.4 is an "OR" gate which produces the binary value 1 as output if any of its inputs is 1, with the result being the binary value 0 if both inputs are 0.

Finally, a NOT gate (Fig. 13.5) accepts only a single input which it reverses. That is, if the input is "1," the value "0" is produced and vice versa.

Fig. 13.3 Binary AND operation

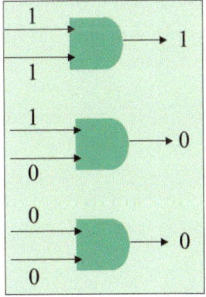

Fig. 13.4 Binary OR
operation

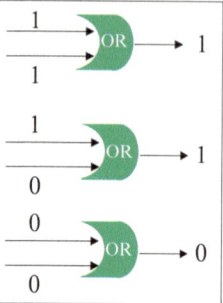

Fig. 13.5 NOT operation

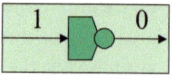

Fig. 13.6 Half adder

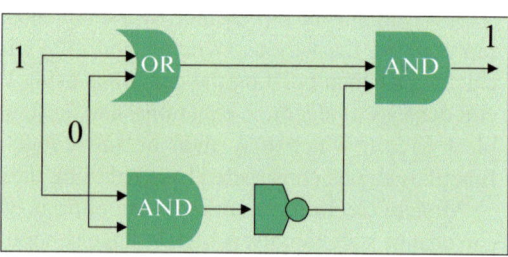

The logic gates may be combined to form more complex circuits. The example in Fig. 13.6 is that of a half adder of 1 + 0. The inputs to the top OR gate are 1 and 0 which yields the result of 1. The inputs to the bottom AND gate are 1 and 0 which yield the result 0, which is then inverted through the NOT gate to yield binary 1. Finally, the last AND gate receives two 1's as input, and the binary value 1 is the result of the addition. The half adder computes the addition of two arbitrary binary digits, but it does not calculate the carry. It may be extended to a full adder that provides a carry for addition.

Chapter 14
C and C++ Programming Languages

Dennis MacAlistair Ritchie was an American computer scientist who developed the C programming language, and he co-developed the Unix operating system with Ken Thompson at Bell Labs in the early 1970s. C influenced later language development, including C++, which is an object-oriented extension to C. C++ was developed by Bjarne Stroustrup at Bell Labs.

Ritchie was born in New York in 1941 and he joined Bell Labs in 1967. He was involved in the Multics operating system project, and he designed and implemented the C programming language in the early 1970s. The origin of this language is closely linked to the development of the Unix operating system, and C was originally used for systems programming. It later became popular for both systems and application programming, and it influenced the development of other languages such as C++ and Java.

Richie and Thompson received the ACM Turing Award in 1983, in recognition of their achievements in the implementation of the Unix operating system. They received the National Medal in Technology and Innovation from Bill Clinton in 1999 (Fig. 14.1). Stroustrup received the Grace Murray Hopper Award for the development of C++ in 1993.

14.1 C Programming Language

Ritchie developed the C programming language at Bell Labs in the late 1960s/early 1970s, and it became a popular general-purpose programming language. It is used for both systems programming and in application development and is widely used in industry. It was standardized by the American National Standards Institute (ANSI) in 1989, and it became the ISO/IEC 9899 standard in 1990 (the latest version is ISO/IEC 9899:2011).

The language was originally designed to write the kernel for the Unix operating system. Unix was initially written in assembly language for the PDP-7 minicomputer

© Springer Nature Switzerland AG 2018
G. O'Regan, *The Innovation in Computing Companion*,
https://doi.org/10.1007/978-3-030-02619-6_14

Fig. 14.1   Ken Thompson and Dennis Ritchie with President Clinton in 1999

(it had been traditional up to then to write the operating system kernel in an assembly language). Unix was then rewritten in C and ported to the PDP-11 minicomputer, and this was one of the earliest uses of a high-level language for writing the operating system kernel (the earlier Multics operating was written in the PL/1 high-level programming language).

The successful use of C in writing the Unix kernel led to its use as a systems programming language on several other operating systems (e.g., Windows and Linux). C also influenced later language development including C++, Java, C#, Perl, and Unix's C shell. The language is described in detail in (Kernighan and Ritchie 1978).

The language was influenced by the *Basic Combined Programming Language* (BCPL) designed by Martin Richards at the University of Cambridge in the mid-1960s. BCPL was the first brace programming language (i.e., it employed "{" and "}" to represent a block of source code statements). BCPL had only one data type (a word of a fixed number of bits usually chosen to represent the architecture of the underlying machines and capable of representing any valid storage address). For most machines at the time, the word length was 16 bits, but this later posed problems when BCPL was used on machines where the smallest word was 8 bits, 32 bits, or 64 bits.

Ken Thompson developed a simplified version of BCPL, and the language was christened B. However, B's inability to deal with byte addressability led to the C programming language (The name C was chosen as it is the next letter in the alphabet after B, and C++ is the successor of C).

The C programming language provides high-level and low-level capabilities, and the language is portable in that a C program written in ANSI C may be compiled for a very wide variety of computer platforms and operating systems (with minimal changes to the source code).

C is a procedural programming language, and it provides low-level access to the computer memory. It includes conditional statements such as the *if statement* and the *switch statement*; iterative statements such as the *while* statement, the *for* statement, and the *do* statement; and the *assignment* statement.

- If statement

```
if (A == B)¹
        A =  A + 1;
else
        A = A - 1;²
```

- Assignment statement

```
i  = i + 1;
```

One of the first programs that people write in C is the *Hello world* program, which is given by:

```
#include <stdio.h>
int main()
{
        printf("Hello, World\n");
}
```

C includes a standard i/o library (stdio.h), and it has several predefined data types including integers and floating-point numbers:

```
int             (integer)
long            (long integer)
float           (floating point real)
double          (double precision real)
```

C allows more complex data types to be created using *structs* (these are like records in Pascal), and each struct contains several data elements. C supports the use of pointers to access memory locations, and this allows the memory locations to be directly referenced and modified. For example, the result of the following fragment

---

[1] One common error in C programs is writing "=" instead of "==." This totally alters the meaning of the statement.

[2] The semicolon in Pascal is used as a statement separator, whereas it is used as a statement terminator in C.

is to assign the value 5 to the variable *x* (where the memory address of variable *x* is given by &*x*).

```
int     x;
int     *ptr_x;

x = 4;
 ptr_x = &x;
 *ptr_x =5;
```

C is a block-structured language, and a program is structured into functions (or blocks). Each function block contains variables and functions, and a function may call itself (i.e., *recursion is allowed*).

```
if (a == b)
        a++;                        .... Program fragment A
   else
        a--
```

```
if (a = b)
        a++;                        .... Program fragment B
   else
        a--
```

One key criticism of C is that it is easy to make errors in C programs and to thereby produce undesirable results. For example, one of the easiest mistakes to make is to accidentally write the assignment operator (=) for the equality operator (==). This totally changes the meaning of the original statement, as can be seen in program fragments A and B above.

Both program fragments are syntactically correct, and the intended meaning of a program has been changed by writing "=" instead of "==." The philosophy of C to allow statements to be written as concisely as possible, and this is potentially dangerous[3]. The use of pointers potentially leads to problems as uninitialized pointers may point anywhere in memory, and the program may potentially overwrite anywhere in memory.

Therefore, the effective use of C requires experienced programmers, well-documented source code, and formal peer reviews of the source code by other team members.

Brian Kernighan wrote the first tutorial on C, and Kernighan and Ritchie later wrote the popular book *The C Programming Language* (Kernighan and Ritchie 1978). Ritchie later became head of Lucent Technologies Systems Software

---

[3] It is easy to write a one line C program that is incomprehensible. The maintenance of poorly written code is a challenge unless programmers follow good programming practice. This discipline needs to be enforced by formal reviews of the source code.

Research Department (AT&T spun off Bell Labs into a new company called Lucent Technologies).

14.2   C++ Programming Language

Bjarne Stroustrup is a Danish computer scientist who designed and developed the C++ programming language, which is an object-oriented extension of the C programming language. Today, C++ is a widely used general-purpose programming language (Fig. 14.2).

Stroustrup was born in Aarhus, Denmark, in 1950, and he joined the computer science research center at Bell Labs after completing his PhD in the late 1970s. He began analyzing the Unix kernel with respect to distributed computing, and he used his knowledge of Simula 67 (from his PhD work) to extend C with classes and other object-oriented features. His "C with classes" language was appropriately renamed to C++ in 1983, and the first commercial compiler for the language appeared in 1985.

C++ is an object-oriented extension of the C programming language. It was designed to use the power of object-oriented programming and to maintain the speed and portability of C. It provides a significant extension of C's capabilities, but it does not force the programmer to use the object-oriented features of the language.

Fig. 14.2 Bjarne Stroustrup

C++ inherits much of its syntax from C, and the C++ equivalent of the "Hello World" program is given by:

```
#include <iostream>
int main()
{
    std::cout << "Hello World \n";
}
```

A key difference between C++ and C is in the concept of a class. A *class* is an extension to the concept of a structure, which is used in C. The main difference is that whereas a C data *structure* can hold only *data*, a C++ *class* may hold both *data* and *functions*. An *object* is an instantiation of a class: i.e., the class is essentially the type, whereas an object is essentially a variable of that type. Classes are defined in C++ by using the keyword *class* as follows:

```
class class_name
{
    access_specifier_1:
            member1;
    access_specifier_2:
            member2;
    ...
}
```

The members may be either data or function declarations, and an access specifier is used to specify the access rights for each member (e.g., *private, public,* or *protected*). Private members of a class are accessible only by other members of the same class; public members are accessible from anywhere where the object is visible; and protected members are accessible by other members of same class and from members of their derived classes. The following is an example of the class of a rectangle.

```
class CRectangle
{
    int x, y;
      public:
            void set_values (int,int);
            int area (void);
    } rect;
```

C++ provides the standard object-oriented features such as abstraction, encapsulation, inheritance, and polymorphism. Stroustrup published the C++ programming language in 1985 (Stroustrup 2013), and this remained the de facto standard for the language until the publication of the ISO/IEC 14882 standard in 1998.

Chapter 15
Cloud Computing and Distributed Systems

Cloud computing is a type of Internet-based computing that provides computing processing resources on demand. It provides access to a shared pool of configurable computing resources such as networks, servers, and applications on demand, and such resources may be provided and released with minimal effort. It provides users and organizations with capabilities to store and process their data in third-party data centers that may be in distant geographical locations.

A key advantage of cloud computing is that it allows companies to avoid large up-front infrastructure costs such as purchasing hardware and servers, and it also allows organizations to focus on their core business rather than the physical infrastructure. Further, it allows companies to get their applications operational in a shorter time period, as well as providing an efficient way for companies to adjust resources to deal with fluctuating demand. Companies can scale up as computing needs increase and scale down as demand decreases. Cloud providers generally use a "pay as you go" model (Fig. 15.1).

Among the well-known cloud computing platforms are Amazon's Elastic Compute Cloud, Microsoft's Azure, and Oracle's Cloud. The main enabling technology for cloud computing is *virtualization*, which separates a physical computing device into one or more virtual devices. Each of the virtual devices may be easily used and managed to perform computing tasks, and this leads to the creation of a scalable system of multiple independent computing devices that allows the idle physical resources to be allocated and used more effectively.

Cloud computing providers offer their services according to different models. These include *infrastructure as a service* (IaaS) where computing infrastructure such as virtual machines and other resources are provided as a service to subscribers. *Platform as a service* (Paas) provides capability to the consumer to deploy infrastructure related or application related that are supported by the provider onto the cloud. PaaS vendors offer a development platform to application developers.

Software as a service (SaaS) provides capability to the consumer to use the provider's applications running on a cloud infrastructure through a web browser or a

© Springer Nature Switzerland AG 2018
G. O'Regan, *The Innovation in Computing Companion*,
https://doi.org/10.1007/978-3-030-02619-6_15

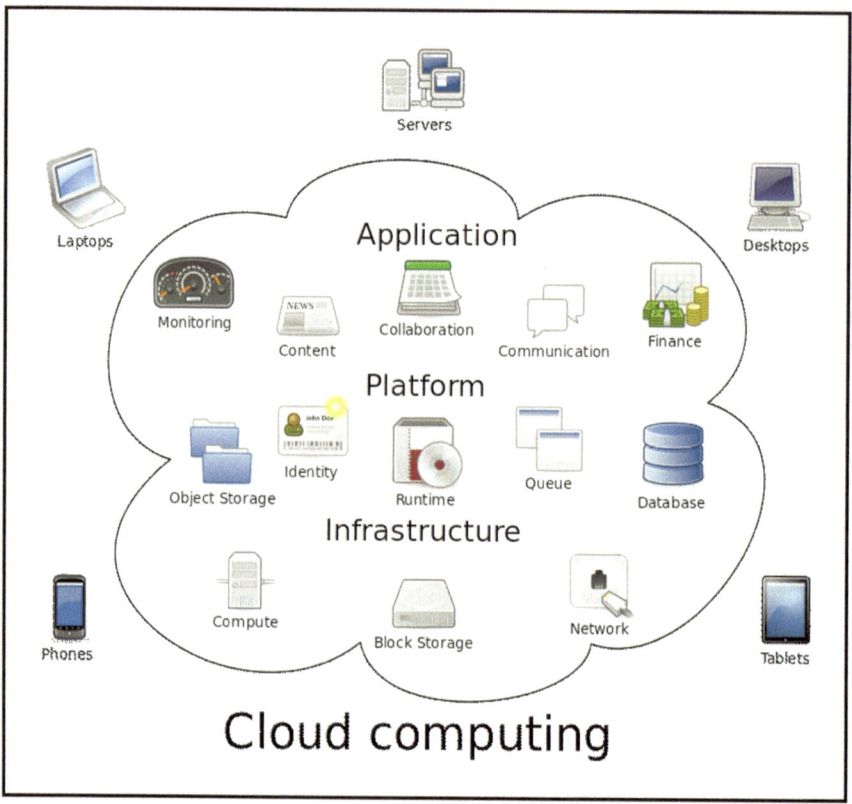

Fig. 15.1 Cloud computing. Creative commons

program interface. Cloud providers manage the infrastructure and platforms that run
the applications.

15.1   Software as a Service

The idea of software as a service (SaaS) is that the software may be hosted remotely
on a server (or servers) and access is provided to it over the Internet through a web
browser. The functionality is provided at the remote server with client access provided through the web browser.

The cost model for traditional software is made up of an up-front cost for a per-
petual license and optional ongoing support fees. SaaS is a software licensing and
delivery model where the software is licensed to the user on a subscription basis.
The software provider owns and provides the service, whereas the user of the service pays a subscription for its use. In some instances, the software may be free to

use with funding provided through advertisements, or there may be a free basic service provided with charges applied for the more advanced version.

A key benefit of SaaS is that the cost of hosting and management of the service is transferred to the service provider, with the provider responsible for resolving defects and installing upgrades of the software. Consequently, the initial setup costs for users are significantly less than for traditional software.

The disadvantages to the user are that data is transferred at the speed of the network, and so the transfer of a large amount of data may take a lot of time. The subscription charges may be monthly or annual, with extra charges possibly applied depending on the amount of data transferred.

15.2 Service-Oriented Architecture

Service-oriented architecture (SOA) is a way of developing a distributed system using stand-alone web services executing on distributed computers in different geographic regions. It is an approach to creating an architecture based upon the use of services, where a service may carry out some small function such as producing data or validating a customer.

A *web service* is a computational or information resource that may be used by another program and it allows the service provider to provide a service to an application (*service requestor*) that wishes to use the service. The web service may be accessed remotely and is acted upon independently. The *service provider* is responsible for designing and implementing the services and specifying the interface to the service.

The service is platform and implementation language independent, and the service provider designs and implements it and specifies the interface. Information about the service is then published in an accessible registry, and service clients (requestor) can locate the service provider and link their application with the specific service and communicate with it. The idea of a SOA is illustrated in Fig. 15.2 below:

There are several standards that support communication between services, as well as standards for service interface definition (Sommerville 2011).

Fig. 15.2 Service-oriented architecture

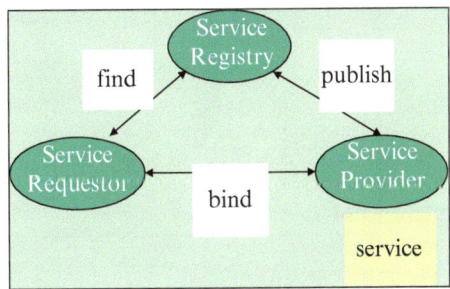

15.3   Distributed Systems

A distributed system (Fig. 15.3) is a collection of computers, interconnected via a network, which is capable of collaborating on a task. It appears to be a single integrated computing system to the user, and most large computer systems today are distributed systems. The components (or nodes) of a distributed system are located on the networked computers, and they interact to achieve a common goal.

A distributed system is not centrally controlled, and therefore individual computers may behave differently at different times, and each computer has a limited and incomplete view of the system. The communication and coordination of action are via message passing.

A distributed system allows hardware and software resources (e.g., printers and files) to be shared, and information may be shared between people and processes located in distant geographical regions. It supports concurrency with multiple processors running on different computers on the network, with the processors running concurrently in parallel, and each computer is running its own local operating system.

A distributed system is designed to tolerate failures on individual computers and to remain in service when a node fails. That is, a distributed system is designed for fault tolerance and to remain operational when there are hardware, software, or network failures. This requires recovery and redundancy features (such as the duplication of information on several computers) to be built in, with continuity of service (possibly a degraded service) when failures occur.

The design of a distributed system is more complex than a centralized system, as there may be complex interactions between its components and the system infrastructure. Its performance is dependent on the network bandwidth and load, as well on the speed of the computers that are on the network. This differs from a centralized

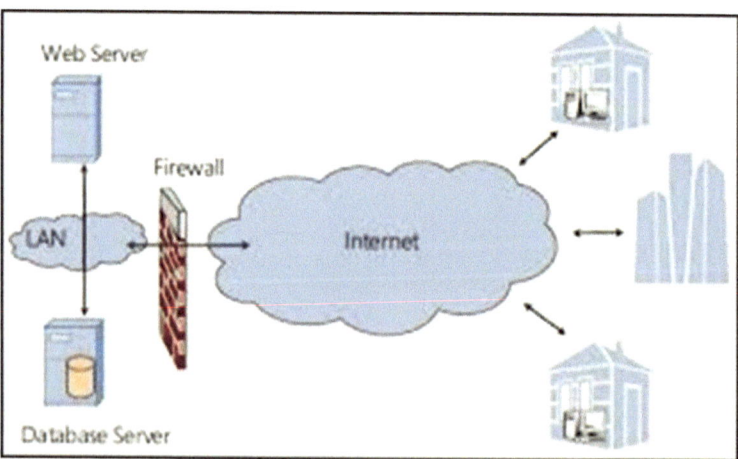

Fig. 15.3 A distributed system

system, which is dependent on the speed of a single processor. The performance and response time of a distributed system may vary (and be unpredictable) depending on the network load and network bandwidth, and the response time may vary from user to user.

The nodes in a distributed system are often independent systems with no central control, and the network connecting the nodes is a complex system, which is not controlled by the computers using the network. There are many applications of distributed systems such as fixed line, mobile and wireless networks, company intranets, the Internet, and the World Wide Web.

Chapter 16
CMMI and Software Process Improvement

Software process improvement is concerned with practical action to improve the processes in the organization, to ensure that they meet business goals more effectively. For example, the business goal might be to develop and deliver high-quality software products faster to the market place, and so the associated processes need to be improved to achieve this. The origins of the software process improvement field go back to the manufacturing sector and to Walter Shewhart's work on statistical process control in the 1930s.

Shewhart's work was later refined by Deming and Juran, who argued that high-quality processes are essential to the delivery of a high-quality product. They argued that the quality of the product is largely determined by the processes used to produce and support it. Therefore, there is a need to focus on the process as well as the product itself, as high-quality products are built from high-quality processes. Deming and Juran's approach transformed struggling manufacturing companies with quality problems, to companies that could consistently produce high-quality products. Their approach enabled companies to achieve cost reductions and higher productivity, as less time was spent in reworking defective products (O'Regan 2014).

Deming and Juran's work was later applied to the software quality field by Watt Humphries and others at the *Software Engineering Institute* (SEI). This led to the birth of the software process improvement field in the late 1980s, and Watt Humphries is considered the father of software quality. Humphries published the book "*Managing the Software Process*" (Humphry 1989) in 1989, and he asked fundamental questions such as:

- How good is the current software process?
- What must I do to improve it?
- Where do I start?

Software process improvement initiatives support the organization in achieving its key business goals such as delivering software faster to the market, improving quality, and reducing or eliminating waste. The goal is to work smarter to build soft-

© Springer Nature Switzerland AG 2018
G. O'Regan, *The Innovation in Computing Companion*,
https://doi.org/10.1007/978-3-030-02619-6_16

ware better, faster, and cheaper than competitors. Software process improvement makes business sense, and it provides a tangible return on the investment made.

Humphries recognized that a software process improvement initiative involves change to the way that work is done, and it therefore needs top management support to be successful. It requires the participation of the software engineers, and changes to the processes are made based on an understanding of their strengths and weaknesses. Every task and activity can be improved, and change is continuous. The new processes need to be reinforced with training, and independent audits are conducted to ensure process fidelity.

The Software Engineering Institute[1] developed the Capability Maturity Model (CMM®) in the early 1990s as a framework to help software organizations improve their software process maturity. The CMMI is the successor to the older CMM, and its implementation brings best practice in software and systems engineering into the organization. The SEI and many other quality experts believe that there is a close relationship between the maturity of software processes and the quality of the delivered software product. The first version of the CMM was released in 1991, and the first version of the CMMI was released in 2001.

The CMM built upon the work of quality gurus such as Deming and Juran, who transformed companies with major quality problems by improving their manufacturing processes. Similarly, software companies need good software processes to deliver high-quality software, and the SEI has collected empirical data to show that there is a close relationship between software process maturity and the quality of the software. Therefore, there is a need to focus on the software process as well as on the product itself.

The CMMI is a framework to assist in the implementation of best practice in software and systems engineering. It is an internationally recognized model for process improvement and is used worldwide by thousands of organizations. The focus of the CMMI is on the improvements to the software process to ensure that they meet business needs more effectively.

A *process* is a set of practices or tasks performed to achieve a given purpose. It may include tools, methods, material, and people. An organization will typically have many processes in place for doing its work, and the object of process improvement is to improve these to meet business goals more effectively.

The process is an abstraction of the way in which work is done in the organization and is the glue (Fig. 16.1) that ties people, procedures, and tools together.

The process may be described by a process map which details the flow of activities and tasks. The process map will include the input to each activity and the output from each activity. Often, the output from one activity will become the input to the next activity. As a process matures, it is defined in more detail and documented. It

---

[1]The SEI was founded by the US Congress in 1984, and it has been successful in advancing software engineering practices in the United States and worldwide. It performs research to find solutions to key software engineering problems, and its solutions are validated through pilots. These are then disseminated to the wider software engineering community through its training program.

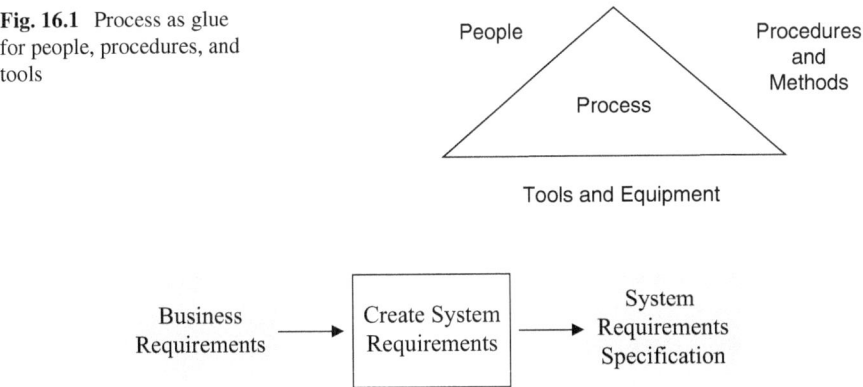

Fig. 16.1 Process as glue for people, procedures, and tools

Fig. 16.2 Sample process map

will have clearly defined entry and exit criteria, inputs and outputs, an explicit description of the tasks, verification of the process, and consistent implementation throughout the organization. A simple example of a process map for creating the system requirements specification is described in Fig. 16.2.

The CMMI process model has proved to be effective in assisting companies in improving their software engineering practices and in achieving consistent results in the delivery of high-quality software. The SEI maintains data on the benefits gained by organizations from their use of the CMM and CMMI. Table 16.1 presents a summary of the improvements gained by 25 organizations (Software Engineering Institute 2006).

For example, *Northrop Grumman Defense Systems* met every milestone (25 in a row) with high quality and customer satisfaction; *Lockheed Martin* reported an 80% increase in software productivity over a 5-year period when it achieved CMM level 5. *Siemens (India)* reported an improved defect removal rate from over 50% before testing to over 70% before testing and a post-release defect rate of 0.35 defects per KLOC. *Accenture* reported a 5:1 return on investment from software process improvement activities.

A process model[2] such as the CMMI defines best practice for software processes in an organization. It describes what the processes should do rather than how they should be done, which allows the organization to use its professional judgment in the implementation of the processes. The process model is interpreted and tailored to meet the needs of the organization.

A process model provides a place to start an improvement initiative, and it provides a common language and shared vision for improvement. It provides a framework to prioritize actions, and it allows the benefits of the experience of other

---

[2] There is the well-known adage, "All models are wrong, some are useful."

Table 16.1 Benefits of software process improvement (CMMI)

Improvements	Median	#Data points	Low	High
Cost	20%	21	3%	87%
Schedule	37%	19	2%	90%
Productivity	62%	17	9%	255%
Quality	50%	20	7%	132%
Customer satisfaction	14%	6	−4%	55%
ROI	4.7:1	16	2:1	27: 1

organizations to be shared. There are several popular process models used in software process improvement including:

- Capability Maturity Model Integration (CMMI)
- ISO 9001 Standard
- ISO 15504
- PSP and TSP
- Six Sigma

Software process improvement provides a return on investment and makes business sense (O'Regan 2010).

16.1   Capability Maturity Model Integrated (CMMI)

The CMMI helps companies to deliver high-quality software systems on time and on budget. It is an internationally recognized model for software process improvement and is used in thousands of organizations around the world. It enables management to identify the processes that it needs to improve and what it should do to improve them. It is used to implement best practice in software and systems engineering.

The CMMI is a process model, and it defines the characteristics or best practices of good processes. It does not prescribe how the processes should be done, and it gives the organization the freedom to interpret the model to suit its context and business needs. It provides a roadmap for an organization to get from where it is today to a higher level of maturity. The advantage of model-based improvement is that it provides a place to start process improvement, as well as a common language and a shared vision.

The historical CMM consisted of five maturity levels with the higher maturity levels representing advanced software engineering capability. The lowest maturity level is level 1, and the highest is level 5. The SEI developed an assessment methodology (CBA IPI) to determine the maturity of software organizations, and initially most organizations were assessed at level 1 maturity. However, over time companies embarked on improvement initiatives and matured their software processes, and today many companies are performing at the higher maturity levels.

The first company to be assessed at CMM level 5[3] was the Motorola plant in India. The success of the software CMM led to the development of other process maturity models such as the systems engineering capability maturity mode (CMM/ SE) which is concerned with maturing systems engineering practices and the people capability maturity model (P-CMM) which is a framework for HR maturity.

The development of the CMMI® (Chrissis et al. 2011) began in the late 1990s, and it involved merging the software CMM and the systems CMM. It is compatible with ISO 15504[4], and it consists of five maturity levels, with each maturity level (except level 1) consisting of several process areas. Each process area consists of a set of goals, which must be implemented by practices related to that process area, and processes need to be defined and documented. The users of the process need to receive appropriate training to enable them to carry out the process, and processes are enforced with independent audits.

The emphasis on level 2 of the CMMI is on maturing management practices such as project management, requirements management, configuration management, and so on. The emphasis on level 3 of the CMMI is to mature engineering and organization practices, including peer reviews and testing, requirements development, software design, and so on. Level 4 is concerned with ensuring that key processes are performing within strict quantitative limit, and adjusting processes, to ensure performance is within these limits. Level 5 is concerned with continuous process improvement that is quantitatively verified.

Maturity levels may not be skipped in the *staged implementation* of the CMMI. There is also a *continuous representation* of the CMMI that allows the organization to focus on improvements to key selected processes. However, in practice, it is often necessary to implement several of the level 2 process areas before serious work can be done on implementing a process at a higher maturity level. The use of metrics becomes more important as an organization matures, as metrics allow the performance of an organization to be objectively judged.

For more detailed information on software process improvement, see (O'Regan 2010, 2014).

---

[3] However, the fact that a company has been appraised at a certain CMM or CMMI rating is no guarantee that it is performing effectively as a business. For example, the Motorola plant in India was assessed at CMM level 5 in the late 1990s, while Motorola fell behind in the GSM market.

[4] ISO 15504 (popularly known as SPICE) is an international standard for software process assessment. There are dedicated versions of SPICE for the automotive and medical device sectors.

Chapter 17
Colossus and Code Breaking at Bletchley Park

The team at Bletchley Park in England played an important role in breaking the Enigma and Lorenz codes during the Second World War, and their code-breaking work contributed to the ultimate defeat of Nazi Germany. The Enigma codes were used by the Germans in the transmission of naval messages to their submarines in the Atlantic and were designed to ensure that any unauthorized interception of the messages would be unintelligible to a third party (Fig. 17.1).

The Enigma machine converted the plaintext (i.e., the original unencrypted message) into the encrypted text, and these encrypted messages were then transmitted by the Germans to their submarines in the Atlantic or to their bases throughout Europe. Any unauthorized interception of these messages yielded no information, as the encrypted messages were meaningless to the interceptor.

A team of cryptanalysts at Bletchley Park cracked the Enigma codes. They developed a code-breaking machine called the Bombe, and this machine found potential settings for the Enigma machine on that day. The Poles had done some work on code breaking prior to the start of the war, and they passed their knowledge on to the British after the German invasion of Poland.

Alan Turing and Gordon Welchman built on the Polish research to develop the *bombe machine*. They observed that a message often contained common words or phrases, such as a general's name or weather reports. This enabled them to guess short parts of the original message. These guesses were called *cribs* (Fig. 17.2).

The Enigma machine did not allow a letter to be enciphered to itself, and this reduced the potential number of settings that the Enigma machine could be in on that day. The code-breaking team then wired the bombe to check the reduced set of settings. The bombe found potential Enigma settings not by proving a setting but by disproving every incorrect one in turn. The bombe machine played an important role in the Battle of the Atlantic and in protecting British and Allied shipping (Fig. 17.3).

© Springer Nature Switzerland AG 2018

G. O'Regan, *The Innovation in Computing Companion*,

https://doi.org/10.1007/978-3-030-02619-6_17

Fig. 17.1 Bletchley Park

Fig 17.2 The Enigma
machine. Public Domain

The first bombe was installed in early 1940, and there were over 200 machines in operation by the end of the war. The British Tabulating Machine Company built them, and a replica bombe was rebuilt at Bletchley Park by a team of volunteers in 2008.

Tommy Flowers made important contributions to breaking the *Lorenz codes* during the Second World War. He led the team that designed and built Colossus, which

Fig 17.3 Rebuilt Bombe. Photo Public Domain

was one of the earliest electronic computers. The machine was designed to decode top-level encrypted German military communication. It provided British and American Intelligence with vital information on German military plans around the D-Day invasion and later battles and helped to ensure the success of the Normandy landings and the defeat of Nazi Germany.

Alan Turing introduced Flowers to Max Newman who was leading British efforts to break a German cipher generated by the Lorenz SZ42 machine. This teletype-writer coding machine was considerably more complex than the Enigma machine, and Turing's introduction led to Flowers becoming involved with the code-breaking work at Bletchley Park.

The Lorenz codes were used in the transmission of important messages between the German High Command in Berlin and their military commanders in the field. The machine was based on the *Vernam cipher*, and the Lorenz SZ 40/42 machine performed the encryption. The Bletchley Park code breakers called the machine *Tunny* and the coded messages *Fish*.

The initial approach used for deciphering the Lorenz codes was with the Heath Robinson machine (a slow and unreliable machine). Flowers proposed an alternate solution involving the use of an electronic machine in 1943. This machine was called Colossus, and it employed 1800 thermionic valves. The management at Bletchley Park were skeptical but encouraged him to continue with his work.

Flowers and others at the Post Office Research Center built the machine at the Post Office Research Station at Dollis Hill in London in 11 months. Its successor, the Colossus Mark 2, contained 2400 valves and commenced operations on June 1, 1944. It was a large bulky machine and took up the space of a small room and weighed a ton.

The machine was used to find possible key combinations for the Lorenz machines rather than decrypting an intercepted message in its entirety. The code-breaking work involved carrying out complex statistical analyses on the intercepted messages.

It compared two data streams to identify possible key settings for the Lorenz machine. The first data stream was the encrypted message, and it was read at high speed from a paper tape. The second stream was generated internally and was an electronic simulation of the Lorenz machine at various trial settings. If the match count for a setting was above a certain threshold, it would be sent as output to an electric typewriter.

It provided vital information for the Normandy landings and confirmed that Hitler had been successfully misled by Allied disinformation into believing that the Normandy landings were to be a diversionary tactic. Further, it confirmed that no additional German troops were to be moved there. The Colossus Mark 2 machine helped the Allies in monitoring the German reaction to their deception tactics.

The Colossus Mark 1 machine was specifically designed for code breaking rather than as a general-purpose computer. It was semi-programmable and helped in deciphering messages encrypted using the Lorenz machine. A prototype was available in 1943, and a working version was available at Bletchley Park in early 1944. The Colossus Mark 2 was introduced just prior to the Normandy landings (Fig. 17.4).

The Colossus Mark 1 used 15 kW of power and could process 5000 characters of paper tape per second. It enabled a large amount of mathematical work to be done in hours rather than in weeks. There were ten Colossi machines working at Bletchley Park by the end of the war. A replica of the Colossus was rebuilt by a team of volunteers led by Tony Sale from 1993 to 1996 and is at Bletchley Park Museum.

Fig 17.4 Colossus Mark 2. Photo Public Domain

The contribution of Bletchley Park to the cracking of the German Enigma and Lorenz codes and to the early computing field remained clouded in secrecy until recent times. The museum at Bletchley Park provides insight into its important contributions during the Second World War, and more information is available in McKay (2011).

17.1   Vernam Cipher

Gilbert Vernam was an AT&T research engineer who invented the Vernam cipher. He invented a stream cipher in 1917 and co-invented the *one-time pad* cipher with Joseph Mauborgne of the US Army Signal Corps. Claude Shannon later showed that if the one-time pad cipher is correctly implemented, then it is theoretically unbreakable and that any unbreakable system essentially has the same characteristics as the one-time pad. The key must be kept secret and needs to be truly random and of the same size as the plaintext. Further, the key must never be reused (either part or whole): i.e., the key must be used only once.

The stream cipher proposed by Vernam involved using a previously prepared key kept on a paper tape to combine character by character with the plaintext message to form the ciphertext. The deciphering of the message involved using the same key combining character by character with the ciphertext to form the plaintext.

Vernam's encryption system used conventional telegraphy practice with the paper tape of the plaintext combined with the paper tape of the key. The intention was that each key tape would be unique, but there were practical problems with generating and distributing such tapes. This problem was solved in the 1920s with the invention of rotor cipher machines to produce a key stream to act instead of a tape. The Lorenz SZ40/42 was one of these.

The encryption and decryption are defined by addition modulo 2, and it is symmetric with the same key employed for both encryption and decryption. It is defined by

$$\text{Ciphertext} = \text{Plaintext} \oplus \text{Key}$$
$$\text{Plaintext} = \text{Ciphertext} \oplus \text{Key}$$

where the $\oplus$ symbol means that a logical or (XOR) operation is performed.

If the key stream is truly random and is used only once, then this is effectively a one-time pad. *However, if the key stream is used for two messages, then the effect of the key stream can be eliminated*, and it is possible for cryptanalysts to derive the message by linguistic cryptanalytical techniques.

$$\text{Ciphertext}_1 \oplus \text{Ciphertext}_2 = \text{Plaintext}_1 \oplus \text{Plaintext}_2$$

It was operator error of this sort that enabled the British to crack the Lorenz codes.

Chapter 18
Commodore PET and 64 Computers

Commodore Business Machine was founded by Jack Tramiel in 1955, and it later became a leading American home computer and electronic manufacturing company. It played an important role in the development of the home computer industry in the 1970s and 1980s, and it is especially remembered for its Commodore PET computer (which was very popular in the education field) and its Commodore 64 home computer.

Tramiel set up Commodore as a typewriter repair business in New York in the mid-1950s, and it diversified into the manufacture of mechanical calculators from the early 1960s. It began manufacturing electronic calculators, including both consumer and scientific calculators, from the late 1960s, and by the early 1970s, it was one of the most popular brands for calculators. The calculators used Texas Instruments chips, but when Texas Instruments entered the calculator market in the mid-1970s, Commodore was unable to compete with the prices offered by Texas.

Commodore purchased a semiconductor company, MOS Technologies, with the intention of using its chips in its calculators. However, Chuck Peddle, one of MOS's employees, convinced Tramiel that the future was in computers and not calculators. Commodore used one of MOS's Technologies chips, the 8-bit 6502, to enter the home computer market in 1977 with the launch of its Commodore Personal Electronic Transactor (PET) computer. This popular computer was mainly used in schools, and one of its models was called the *Teachers' PET*.

Commodore introduced the 8-bit VIC-20 home computer in 1981, and this low-cost machine enabled users to learn about programming and to play video games. Its successor, the Commodore 64, was introduced the following year, and it was a popular machine with good sound and graphics, a color display, 64 Kb of RAM, and Microsoft BASIC.

Commodore purchased a start-up company called Amiga in 1984, and it introduced the 32-bit Amiga 1000 computer in 1985. This machine had advanced graphics and sound and used the Motorola 68000 microprocessor. New and more powerful

© Springer Nature Switzerland AG 2018
G. O'Regan, *The Innovation in Computing Companion*,
https://doi.org/10.1007/978-3-030-02619-6_18

Amiga models were introduced up to the mid-1990s, but ultimately it was the IBM PC and its clones that would dominate the personal computer market. This led to the demise of Commodore in the mid-1990s.

18.1   Commodore PET

Commodore introduced its first computer, the Commodore Personal Electronic Transactor (PET) home computer, in 1977 (Fig. 18.1). This successful home computer was very popular in the education market, and it used the MOS 8-bit 6502 microprocessor, which was designed by Check Peddle and others at MOS Technology. The 6502 controlled the screen, keyboard, cassette recorder, and any peripherals connected to the expansion ports. The machine used the Commodore BASIC operating system. There were several models of the Commodore PET introduced during its lifetime including the PET 2001 series, the PET 4000 series, and the Super PET 8000 series (Fig. 18.1).

The first model introduced was the PET 2001, which had either 4 Kb or 8 Kb of RAM. It had a built-in monochrome monitor with 40 × 25 character graphics enclosed in a metal case. It included a magnetic data storage device known as a datasette (data + cassette) in the front of the machine as well as a small keyboard. There were complaints with respect to the small keyboard, which led to the development of external replacement keyboards.

The PET 4000 series was launched in 1980, and the 4032 model was very successful at schools. Its all-metal construction and all-in-one design made it ideal for the challenges in the classroom. The 4000 series used a larger 12″ monitor and an

Fig 18.1 Commodore PET 2001 home computer

enhanced BASIC 4.0 operating system. Commodore manufactured a successful variant called the *Teachers'* PET.

Commodore introduced the 8000 series, and the last in the series was the SuperPET or SP9000. It used the Motorola 6809 microprocessor, and it provided support for several programming languages such as BASIC, Pascal, COBOL, and FORTRAN.

18.2   Commodore 64

The Commodore 64 (C64) was introduced in 1982, and it was a very successful 8-bit home computer (Fig. 18.2). Its main competitors were the Atari 400 and Atari 800 and the Apple II computer. The cost of the C64 machine was $595, which was significantly less than its rivals, and Commodore cleverly exploited the price difference to rapidly gain market share. Approximately 15 million of the Commodore 64 machines were sold.

It used the MOS 6501 microprocessor and came with 64 kilobytes of RAM. It had 320 × 200 color graphics with 16 colors using the VIC-II graphics chip and the MOS Sound Interface Device (SID) chip. The SID chip was one of the first sound chips to be included in a home computer.

It came with Commodore BASIC, and support for other languages such as Pascal and FORTRAN was also available. Users could also write programs in assembly language to maximize speed and memory use. The Commodore 64's graphics and sound capabilities were quite advanced for the time, and it was very popular for computer games. It dominated the low-end home computer market during the 1980s (Fig. 18.2).

Commodore published detailed technical documentation to assist enthusiastic users to design and develop applications for the C64. This led to the development of over 10,000 commercial applications such as development tools, games, and office productivity applications. Atari was Commodore's main competitor, but it kept its technical information secret.

The C64 included a ROM-based version of the BASIC 2.0 programming language. There was no operating system as such, and instead the kernel was accessed

Fig 18.2 Commodore 64 home computer

via BASIC commands. BASIC did not allow commands for sound or graphics manipulation, and instead the user had to use the "POKE"[1] command to access these chips directly.

The Commodore 64 remained highly popular throughout the 1980s, and it was still being sold up to the early 1990s.

18.3   The Demise of Commodore

Commodore developed the Amiga family of personal computers in the 1980s and 1990s. The Amiga 1000 (or A1000) was released in 1985, and it became popular for its graphical, audio, and multitasking capabilities. The 1994 edition of the Byte magazine Halfhill (1994) described the A1000 as the first multimedia computer, as it was so far ahead of its time with advanced graphics and sound. The Amiga 500 was released in 1987, and it was the best-selling model in the Amiga family.

However, by the late 1980s, the personal computer market was dominated by the IBM personal computer and compatibles, and the performance of the IBM PC and compatibles had caught up with the Amiga family by the early 1990s.

Commodore introduced PC compatible computers in the early 1990s, but these were not very successful, and they failed to make an impact where high-performance graphics and sound were irrelevant. The company began to become unprofitable, and it filed for bankruptcy in 1994. For further information on Commodore, see Bagnall (2012).

---

[1]The BASIC POKE command changes the content of any address in the 16-bit memory range 0–65635 to a byte value (0–255).

Chapter 19
COBOL and Compilers

Eckert-Mauchly Computer Corporation[1] (EMCC) was one of the earliest computer companies, and it was founded by Presper Eckert and John Mauchly in 1947. It pioneered several fundamental computer concepts such as the *stored program*, *subroutines*, *programming languages*, and *compilers*. It developed one of the first commercial computers, the *Universal Automatic Computer* (UNIVAC), for the United States Census Bureau to allow them to process the 1950 census in the United States.

The machine was designed for business and administrative use, rather than for complex scientific calculations. UNIVAC was later used to accurately predict the result of the 1952 presidential election in the United States from a sample of 1% of the population.

EMCC set up a department to develop software applications for UNIVAC, and Grace Murray Hopper (Fig. 19.1) was hired as one of its first programmers in 1949. Hopper played an important role in the development of programming languages, and she made important contributions to the early development of compilers, programming language constructs, data processing, and the COBOL programming language.

Hopper was a pioneer of computer programming, and she had worked on the Harvard Mark I, Mark II, and Mark III computers. She coined the term *computer bug* when she traced an error in the Mark II computer to a moth stuck in one of its relays. The bug was carefully removed and taped to a daily logbook, and the term "bug" is now ubiquitous.

It was very evident to her that programming in binary machine code is tedious and error prone. Machine code consists of writing a string of 0s and 1s, and so it is easy to make mistakes, and it is time-consuming to identify and correct these errors. She believed that the development of a user-friendly language would encourage wider use of computers, and she recognized that libraries of code would also help to reduce errors and duplication of effort.

---

[1] EMCC was later taken over by Remington Rand, which was later taken over by Sperry becoming Sperry Rand, and finally Sperry merged with Burroughs to become Unisys.

© Springer Nature Switzerland AG 2018
G. O'Regan, *The Innovation in Computing Companion*,
https://doi.org/10.1007/978-3-030-02619-6_19

Fig. 19.1 Grace Murray Hopper with UNIVAC

This led to her idea for a compiler that would act as an intermediate program that would translate the program instructions into machine code that could then be understood by the computer. This would allow programmers to employ the friendly and intuitive notation of a high-level programming language, rather than writing tedious and lengthy instructions in binary code.

Her first compiler, the A-O (Arithmetic Language version 0), was developed on the UNIVAC 1 computer and was released in 1952.[2] It was the first compiler to be developed, and the language used symbolic mathematical code to represent binary code combinations. The compiler translated the mathematical notation into machine code.

Hopper had noticed that business data processing customers were uncomfortable with mathematical notation, and she proposed an English-like business language in the mid-1950s. This data processing language was initially called Business Language version 0 (B-0), and it was later called FLOW-MATICs. The language expressed operations using English-like statements, and it influenced the development of COBOL.

She developed a compiler for the FLOW-MATICs language in the late 1950s, and this is considered the first data processing compiler. She contributed to standardizing compilers and to compiler verification.

---

[2] It was more a linker/loader than a compiler.

She recognized the need for a user-friendly programming language that would be easy for business users, and she was the technical adviser to the CODASYL committee that defined the specification of the COBOL language. This was the first business-oriented programming language, and the FLOW-MATICs compiler was employed in its development.

COBOL was introduced in 1960, and Hopper participated in public demonstrations of the first COBOL compiler, and she designed manuals and tools for the language. It provided a degree of machine independence, as the source code was written once, and then compiled into the machine language of the targeted machine.

19.1   COBOL Programming Language

COBOL (Common Business-Oriented Language) was one of the earliest high-level programming languages. It was developed by the CODASYL (Conference/ Committee on Data Systems Languages) committee, whose members were from industry, universities, and the US government. Their goal was to create a common language that would be suitable for business (usually file-oriented) applications. The expectation was that the use of a common language would reduce the cost of programming, as less time would be spent in rewriting applications for different hardware platforms.

The committee defined the specification of COBOL in 1959, and it was published in 1960. The first ANSI standard for the language was published in 1968, and it has been revised on several occasions since then. An object-oriented version of the language was published in 2002. COBOL is still actively used today with many large business applications developed in the language (often several million lines of code). Further, as the cost of investment in developing COBOL applications is quite high (due to their size), they tend to have a long life span (typically 10–30 years) and are often retained even when new technologies or programming languages become popular. There are several billion lines of legacy COBOL code still in use today.

COBOL has a simple English-like syntax, and the language is very verbose (the language is self-documenting). It is not suitable for systems programming or for scientific applications.

COBOL programs are structured into a hierarchy consisting of divisions, sections, paragraphs, sentences, and statements. There are four possible divisions (identification, environment, data, and procedure division), and each division provides key information to the compiler. This includes information about the program, the environment in which the program will run, a description of the data items processed by the program, and a description of the code used to process the data.

There are three data types in COBOL, and these are numeric, alphabetic, and alphanumeric. The *numeric* type consists only of the digits 0–9, the *alphabetic* type consists of letters A–Z and spaces, and *alphanumeric* consists of digits, letters, and special characters. Data types are declared using a level number, a data name, and a picture clause. For example, the following is the definition of the salary data type (the salary is between 0 and 99,999.99 where the V in the definition represents the decimal point position):

01SalaryPIC9(5)V99

The COBOL equivalent of a record or structure is a group item, and the items in a group item are either elementary items or other group items. For example, the definition of the DateOfBirth group item is:

01 DateOfBirth
    02 Day PIC 9(2)
    02 Month PIC 9(2)
    02 Year PIC 9(4)

The following are some basic arithmetic statements in COBOL:

ADD A to B GIVING C
SUBTRACT A FROM B
MULTIPLY A BY C
DIVIDE A INTO B GIVING C

There are many texts on COBOL and for more detailed information, see Stern et al. (2005).

19.2   Compilers

A compiler is a computer program that translates the source code of a program (which is written in a high-level programming language) into the equivalent machine code of the target machine (or into the low-level assembly language). The phases of a typical compiler are described in Fig. 19.2.

The compiler preserves the semantics of the language, rather than defining them, and the meaning of the programming language constructs is given independently by the language designer (e.g., using axiomatic semantics or denotational semantics).

A compiler performs several of the following operations: lexical analysis, preprocessing,[3] parsing, semantic analysis, code generation, and code optimization (Table 19.1).

Hopper coined the term "compiler," and she developed the first compiler (it was more a linker/loader than a compiler), and this was for the A-O (Arithmetic Language version 0) language, which was developed on the UNIVAC 1 computer n 1952. The language used symbolic mathematical code to represent binary code combinations, and the compiler translated the mathematical notation into machine code.

---

[3] The preprocessor is performed on the source code before the next step in the compilation.

Fig. 19.2 Phases in compiler

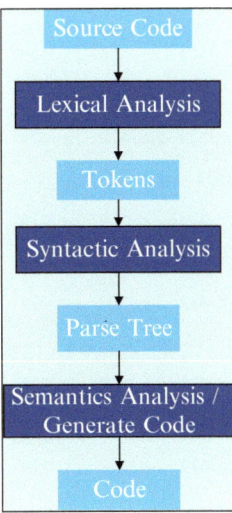

Table 19.1 Compiler operations

Part	Description
Lexical analysis	A lexical analyzer converts a sequence of characters from a computer program into a sequence of tokens (strings with assigned meanings), which are then used by the parser. Lexical analysis is the first phase of compilation
Parsing	Parsing is concerned with the syntactic analysis of the string of symbols in the source code to determine if they conform to the rules of the grammar. It generally involves constructing a parse tree as a structural representation of the input
Semantic analysis	The semantics of a language provides meaning to its constructs. Semantic analysis takes place after parsing and gathers semantic information (e.g., type checking) from the source code
Code generation	Code generation is concerned with converting an intermediate representation of the source code (e.g., parse tree) into a form (e.g., machine code) that can be directly executed on the target computer
Code optimization	Code optimization is concerned with modifying the generated code to make it operate more efficiently

The first complete compiler was developed by John Backus and his colleagues at IBM in 1957, and this was for the FORTRAN programming language. For more information on compilers, see Aho and Ullman (1977).

Chapter 20
Databases

A database (DB) is an organized collection of data that consists of schemas, tables, queries, reports, and views. It holds information about many different types of entities, as well as information about the relationships between them. A computer program (termed the database management system) may easily select and analyze the desired pieces of data.

A *database management system* (DBMS) is a collection of software programs that allows a user to store, modify, and extract data from a database. The interaction between the users and the database is through the DBMS, and it enables the definition, creation, query, update, and administration of databases. There are three main categories of database management systems, namely, the hierarchical, network, and relational models. These differ in how the DBMS organizes data internally, which determines the speed and efficiently of data retrieval from the database.

A *network model* database is perceived by the user to be a collection of record types, with the relationships between the record types organized as a network. A *hierarchical model* is perceived by a user to be a collection of hierarchies or trees, and it is a more restricted structure than the network model as only one arrow may enter each box on the network. A *relational model* is perceived by the user to be a collection of tables (or relations), and it has been the most popular category of databases since the 1980s.

Early work on database management systems began in the 1960s, as part of the Apollo mission to land man on the moon. It was clear that the existing systems were incapable of handling the coordination of the vast amounts of data required for the project. IBM developed the Generalised Update Access Method (GUAM) product in 1964, which evolved into Data Language/1 (DL/1). DL/1 is the data management component of the Information Management System (IMS) hierarchical database, which was one of the earliest database management systems when it was introduced in 1968.

© Springer Nature Switzerland AG 2018
G. O'Regan, *The Innovation in Computing Companion*,
https://doi.org/10.1007/978-3-030-02619-6_20

The CODASYL committee[1] set up a database task group and devised a standard, which became known as the "CODASYL approach." This became the network standard, and it was defined in the late 1960s. The standard was introduced in 1971.

Codd proposed a radically new approach to the management of data with his relational model in 1970, and IBM developed the prototype relational database called System R in the mid-1970s. Commercial relational database systems were introduced from the early 1980s, and today they are ubiquitous.

20.1   Hierarchical and Network Models

A database management system uses the network model if the data relationships are defined in terms of a graph. The relationships are defined in terms of records (a record is a collection of fields, with each field containing one value), which are connected via links. Any given record may have several parent records and several dependent records. Cycles are permitted in the model.

Charles Bachman and the CODASYL committee defined the network model in the late 1960s. General Electric's Integrated Data Store (IDS) and the Integrated Database Management System (IDMS) were based on the network model. Figure 20.1 illustrates a possible network view of suppliers and parts in a simple graph-like structure, where many-to-many relationships may be expressed.

A DBMS uses the hierarchical model if the data relationships are defined in terms of hierarchies (i.e., in a tree like structure). The relationships are simple but inflexible (as they are one to many). The data are defined as records, which are connected to each other through links. Each child record may have only one parent, whereas each parent record may have several children records. The whole tree (starting from the root) needs to be traversed to retrieve data from a hierarchical database. That is, the hierarchical model is a more restricted version of the network model, where no box can have more than one arrow entering the box although several arrows can leave a box.

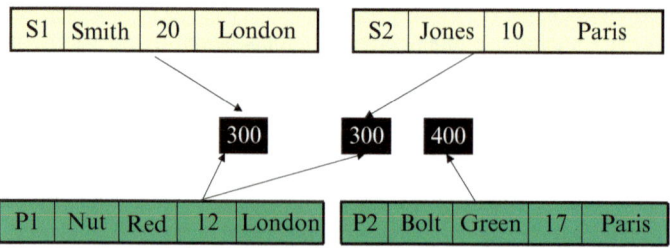

Fig 20.1 Simple part/supplier – network model

---

[1]The CODASYL committee is the group that defined and standardised the COBOL programming language. It was also involved in work in standardizing database interfaces.

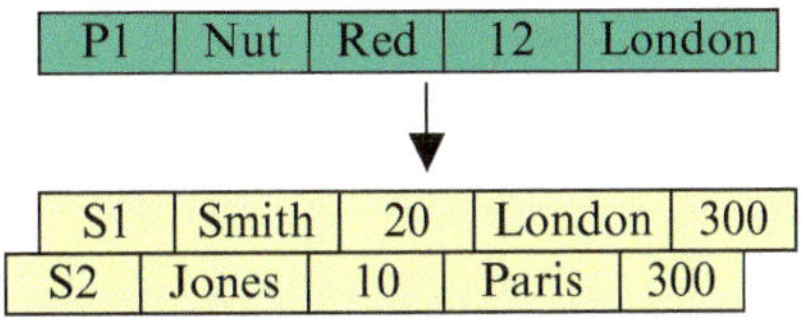

Fig 20.2 Simple part/
supplier – hierarchical
model

P1	Nut	Red	12	London

S1	Smith	20	London	300
S2	Jones	10	Paris	300

Figure 20.2 presents a possible hierarchical view of suppliers and parts in a simple treelike structure. Each tree consists of one part record together with a set of supplier record occurrences, one for each supplier of the part.

The database access and manipulation component of the hierarchical model is termed Data Language/1, and it includes a data definition language and a data manipulation language. The IBM Information Management System (IMS) is a hierarchical database, and it was created in the late 1960s. For more detailed information on the network and hierarchical model, see Date (1981).

20.2   The Relational Model

A relational database management system (RDBMS) is a system that manages data using the relational model, and examples include RDMS developed at MIT in the 1970s, Ingres developed at Berkeley in the mid-1970s, Oracle developed in the late 1970s, DB2, Informix, and Microsoft SQL Server.

A relation is defined as a set of tuples (usually represented by a table). A table is data organized in rows and columns, with the data in each column of the table of the same data type. Constraints may be employed to provide restrictions on the kinds of data that may be stored in the relations, and these are a way of implementing business rules in the database.

Relations have one or more keys associated with them, and the *key uniquely identifies the row of the table*. An *index* is a way of providing fast access to the data in a relational database, as it allows the tuple in a relation to be looked up directly (using the index) rather than checking all tuples in the relation.

The Structured Query Language (SQL) is a computer language that tells the relational database what to retrieve and how to display it. A stored procedure is executable code that is associated with the database, and it is used to perform common operations on the database.

The concept of a relational database was first described in a paper "*A Relational Model of Data for Large Shared Data Banks*" by Codd (1970). A relational database is a database that conforms to the relational model, and it may be defined as a set of relations (or tables).

Codd (Fig. 20.3) was a British mathematician, computer scientist, and IBM researcher, who initially worked on the SSEC (Selective Sequence Electronic Computer) project in New York and then on the IBM 701 and IBM 702 computers

Fig 20.3 Edgar Codd

and the IBM 7030 STRETCH computer (IBM's first transistorized computer). He was the creator of STEM (statistical database expert manager).

He developed the *relational database model* in the late 1960s, and today, this is the standard way that information is organized and retrieved from computers. Relational databases are at the heart of systems from hospitals' patient records to airline flight and schedule information.

IBM was promoting its IMS hierarchical database in the 1970s, and it showed little interest in Codd's new relational database model. However, it agreed to implement Codd's ideas on the *System R research project* in the 1970s, which demonstrated the power of the relational model as well as good transaction processing performance. The project introduced a data query language that was initially called SEQUEL (later renamed to SQL), which was designed to retrieve and manipulate data in the IBM database.

The relational model became popular from the early 1980s, and Codd received the ACM Turing Award in 1981 for its development. A binary relation $R(A,B)$ where A and B are sets is a subset of the Cartesian product $(A \times B)$ of A and B. The domain of the relation is A, and the co-domain of the relation is B. The notation aRb signifies that there is a relation between a and b and that $(a,b) \in R$. An n-ary relation $R(A_1, A_2, \ldots A_n)$ is a subset of the Cartesian product of the n sets, i.e., a subset of $(A_1 \times A_2 \times \ldots \times A_n)$.

The data in the relational model are represented as a mathematical n-ary relation, and so the relation is a set of n-tuples and may be visually represented by a table. The data is organized in rows and columns, and the data stored in each column of the table is of the same data type.

P#	PName	Colour	Weight	City
P1	Nut	Red	12	London
P2	Bolt	Green	17	Paris
P3	Screw	Blue	17	Rome
P4	Screw	Red	14	London
P5	Cam	Blue	12	Paris
P6	Cog	Red	19	London

Fig 20.4 PART relation

The basic relational building block is the domain or data type (often called just type). Each row of the table represents one n-tuple (one tuple) of the relation, and the number of tuples in the relation is the cardinality of the relation. The PART relation in Date (1981) consists of a heading and body and is of cardinality six (Fig. 20.4). There are five data types representing part numbers, part names, part colors, part weights, and locations in which the parts are stored. The body consists of a set of n-tuples.

Strictly speaking there is no ordering defined among the tuples of a relation. However, in practice, relations are often considered to have an ordering. A *normalized relation* satisfies the property that at every row and column position in the table, there is exactly one value (i.e., never a set of values). All relations in a relational database are required to satisfy this condition, and an un-normalized relation may be converted into an equivalent normalized form.

It is often the case that within a given relation, there is one attribute with values that is unique within the relation and can thus be used to identify the tuples of the relation. For example, the attribute P# of the PART relation has this property since each PART tuple contains a distinct P# value, which may be used to distinguish that tuple from all other tuples in the relation. P# is termed the *primary key* for the PART relation. A candidate key that is not the primary key is termed the *alternate key*.

An *index* is a way of providing quicker access to the data in a relational database, as it allows the tuple in a relation to be looked up directly (using the index) rather than checking all the tuples in the relation.

The consistency of a relational database is enforced by a set of constraints that provide restrictions on the kinds of data that may be stored in the relations. The constraints are declared as part of the logical schema and are enforced by the database management system.

Structured Query Language (SQL) is a computer language that tells the relational database what to retrieve and how to display it. It was designed and developed at IBM by Donald Chamberlin and Raymond Boyce, and the SQL operations include *insert*, *delete*, *update*, *query*, schema creation and modification, and data access control. The *data manipulation language* (DML) is the subset of SQL used to add, update, and delete data, and the *data definition language* (DDL) manages table and index structures. A stored procedure is executable code that is associated with the database. It is usually written in an imperative programming language, and it is used to perform common operations on the database. Next, we discuss the Oracle DBMS.

20.3   Oracle Database

Oracle is the world leader in relational database technology, and Larry Ellison (Fig. 20.5) and others founded the company in the late 1970s. Its initial goal was to commercialize relational database technology before IBM, and the release of Oracle V.2 (the first commercial SQL relational database management system) in 1979 was an important milestone in the history of computing.

An Oracle database consists of a large collection of data managed by the Oracle DBMS in a multiuser environment. It allows concurrent access to the data, and it prevents unauthorized access. It provides a smooth recovery of database information in the case of an outage or any other disruptive event.

An Oracle database consists of one or more physical data files, which contain all the database data, and a control file that contains entries that specify the physical structure of the database. A schema is a collection of database objects, and these are the logical structures that directly refer to the database's data. They include structures such as tables, views, and indexes.

Tables are the basic unit of data storage with a *table* having several rows and columns, and an index provides an access path to the table data. A *view* is a customized presentation of data from one or more tables, and it derives data from the actual tables on which it is based. Each Oracle database has a *data dictionary*, which stores information about the logical and physical structure of the database. The data dictionary is created when the database is created and is updated automatically to accurately reflect the status of the database.

A *database administrator* (DBA) is responsible for setting up the Oracle database server and application tools and with allocating system storage and planning future storage requirements. The DBA will create appropriate storage structures to meet the needs of application developers. The access to the database will be monitored and controlled and its performance monitored and optimized. The DBA will plan backups and recovery of database information.

Fig 20.5 Larry Ellison onstage

Chapter 21
DEC PDP-11 and VAX 11/780 Minicomputers

The *minicomputer* was a new class of low-cost computers that arose during the 1960s, and its development was facilitated by the introduction of integrated circuits, which improved performance and reduced costs. Minicomputers were distinguished from the large mainframe computers by price and size, and they formed a class of the smallest general-purpose computers.

Mainframes were large expensive machines (typically costing over $1 million), and they required separate rooms for technicians and operation. Minicomputers often cost under $100,000, and they were designed for direct, personal interaction with the programmer.

Digital Equipment Corporation (DEC) and Control Data Corporation (CDC) introduced small or minicomputers in the early 1960s. These included DEC's PDP-1, which was released in 1961, and the CDC-160A, which was released in 1960. These machines cost $110,000 and $60,000, respectively, which was a fraction of the cost of a mainframe computer.

Ken Olsen and Harlan Anderson founded DEC in 1957 as a spin-off from MIT's Lincoln computer laboratory. It was an innovative and forward-thinking company and became the second largest computer company in the world in the late 1980s. The company had revenues of over $14 billion and over 100,000 employees at its peak. It dominated the minicomputer era from the 1960s to the 1980s, with its PDP and VAX series of computers. These were very popular with the engineering and scientific communities.

DEC's first computer, the *Programmed Data Processor* (PDP-1), was released in 1961. This 18-bit machine was a relatively inexpensive computer for the time, and DEC's minicomputers were relatively affordable to business. The PDP-1 was a simple and reasonably easy-to-use computer with 4000 words of memory, and one of the earliest computer games, Spacewar, was developed for it. There is a restored version of the PDP-1 computer at the Computer History Museum in California.

The PDP-8 minicomputer was released in 1965, and this was a 12-bit machine with a small instruction set. It was a major commercial success for DEC with many sold to schools and universities. The PDP-11 was a highly successful series of 16-bit

© Springer Nature Switzerland AG 2018
G. O'Regan, *The Innovation in Computing Companion*,
https://doi.org/10.1007/978-3-030-02619-6_21

minicomputers, which remained a popular machine for over 20 years from its release in 1970 to the early 1990s.

The PDP series of minicomputers were elegant and reasonably priced, and they dominated the new minicomputer market segment. They were an alternative to the expensive mainframe computers offered by IBM to large corporate customers. Research laboratories, engineering companies, and other organizations with large computing needs all used DEC's minicomputers.

Gordon Bell was one of DEC's earliest employees, and he was the architect of several members of the PDP family (including the PDP-6). He designed the interrupt system and the multiplier/divider unit of the PDP-1 computer, which built upon work done at the MIT Lincoln Laboratory. He played an important role in the development of the PDP family of minicomputers and later became vice-president of research and development at DEC. He led the development of the 32-bit VAX series of computers, and he was involved in the design of around 30 microprocessors.

The VAX series of minicomputers were derived from the best-selling PDP-11. The VAX product line was a competitor to the IBM System/370 series of mainframe computers, and it used the Virtual Memory System (VMS) operating system. The VAX-11/780 was released in 1978, and it was a major success for the company.

The rise of the microprocessor and microcomputer led to the availability of low-cost personal computers, and this later challenged DEC's product line. DEC was slow in recognizing the importance of these developments, and Olsen's statement from the mid-1970s "There is no need for any individual to have a computer in his home" suggests that DEC were totally unprepared for the revolution in personal computing and its threat to DEC's business.

DEC developed its personal computer after the launch of the IBM PC. The DEC machine easily outperformed the PC, but it was more expensive and incompatible with IBM PC hardware and software. DEC's microcomputer efforts were a failure, but its PDP and VAX products were continuing to sell, and by the late 1980s, DEC was threatening IBM's number one spot in the industry. However, the increasing power of the newer generations of microprocessors began to challenge DEC's minicomputer product range.

DEC was too late in responding to the paradigm shift in the industry, and this proved to be fatal for the company. Its sales fell sharply in the early 1990s, and indecision and infighting inside the company delayed an appropriate response.

Robert Palmer became the new CEO with the responsibility to return the company to profitability. He attempted to change the business culture and sold off non-core businesses. Eventually, Compaq acquired DEC in 1998 for $9.8 billion, and HP later acquired Compaq.

21.1   PDP-11 Minicomputer

The PDP-11 (Fig. 21.1) was a family of 16-bit minicomputers produced by DEC from 1970 up to the early 1990s. Harold McFarland designed it, and the prototype was available in 1969. It was released in 1970, and there were several models in the PDP-11 family.

Fig. 21.1 PDP-11

It was one of DEC's most successful computers, with over 600,000 machines sold. It was the only 16-bit computer made by the company, as its successor was the 32-bit VAX:11 series. It started its life as a minicomputer and ended its life as super/microcomputer. The release price of the PDP-11 in 1970 was a very affordable $20,000.

Its central processing unit had eight 16-bit registers, six general-purpose registers, the stack pointer, and a program counter. It included software such as an editor, debugger, and utilities. The size of its memory was 128 KB.

The PDP-11 was useful for multiuser and multitasking applications, and the first version of the UNIX operating system ran on a PDP-11 in 1970. The VAX line at Digital began as an enhancement to the PDP-11 architecture.

21.2   The VAX 11/780 Minicomputer

The Virtual Address eXtension (VAX) was a family of minicomputers produced by DEC from the mid-1970s up to the late 1980s. This family used processors implementing the VAX instruction set architecture, and its members included minicomputers such as the VAX-11/780, VAX-11/782, and so on. The VAX product line was a competitor to the IBM System/370 series of computers.

Fig. 21.2 Vax 11/780

The 32-bit VAX series of computers was introduced following the return of Gordon Bell as the VP of engineering in 1972. The VAX series was derived from the PDP-11, and it was the first widely used 32-bit minicomputer. The VAX 11/780 (Fig. 21.2) was the first member of the family, and it was released in 1978. It was a major success for the company, and it was one of the most successful families of computers of all time. It was the first one MIPS (million instructions per second) machine.

Several programming languages including Fortran 77, BASIC, COBOL, and Pascal were available for the machine. The VAX-11/780 used the DEC VMS (Virtual Memory System) operating system, which was a multiuser, multitasking, and virtual memory operating system. The VAX-11/780 remained the base system that every computer benchmarked its speed against for many years. For more information on DEC, see Schein et al. (2004).

Chapter 22
Digital Photography

Early attempts at photography go back to the early nineteenth century. Joseph Nicéphore Niépce, a French inventor, is credited with its invention in the mid-1820s. Niépce developed a photographic process termed *heliography*, which used a naturally occurring asphalt (Bitumen of Judea) as a coating on glass. This material was light sensitive and hardened on its exposure to light. When the plate was subsequently washed with a solvent, the unhardened parts were washed away, and only the hardened part (the heliographic engraving) remained.

He employed a primitive *camera obscura* (darkroom) to create the oldest surviving photograph of a real-world scene in 1826 (Fig. 22.1). This employed a darkened chamber where a screen with a small hole is used to project the image of the scene through the small hole onto a surface opposite the opening. The image is inverted, and the plate in the figure was exposed to the camera for several hours.

Louis Daguerre (an associate of Niépce) refined the bitumen process, and he experimented with silver-based processes after Niépce's death. He developed the first publicly and commercially available photographic process in 1839 (Fig. 22.2), which remained in use until 1860. The disadvantage of Niépce's photographic process was that it required very long exposures (several hours), and thus it was not really fit for purpose. Daguerre's new process required significantly less exposure time.

Subsequent innovations made photography easier with exposure time reduced to minutes, to seconds, and to fractions of a second. New technologies such as roll films and color photography were developed, and photography took off among the public.

George Eastman pioneered the use of photographic film (initially on paper film and later celluloid), and his first camera called "Kodak" was released in 1888. The commercial development of digital cameras has revolutionized photography, with traditional photochemical methods rarely used today.

© Springer Nature Switzerland AG 2018
G. O'Regan, *The Innovation in Computing Companion*,
https://doi.org/10.1007/978-3-030-02619-6_22

Fig. 22.1 View from the
Window at Le Gras in
1826 – oldest camera
photograph

Fig. 22.2 Daguerreotype
camera 1839

22.1   Digital Cameras

Early work on digital photography commenced in the late 1950s when Russell
Kirsh and others at the National Bureau of Standards[1] (NBS) developed the first
digital image scanner. Kirsh's scanner traced variations in intensity over the surface
of photographs and made the first digital scan. Figure 22.3 shows Kirsh's scan of his
3-month-old son captured in just 30,976 pixels (in 176 ×176 pixels), where the
depth was 1-bit per pixel (i.e., pure black or white with no shades of gray).

NASA sent digital signals of the moon back to earth that were captured with their
space probes while mapping the surface of the moon. Spy satellites also used digital
images during the cold war.

George Smith and Willard Boyle developed the *charge-coupled device* (CCD) at
Bell Labs in 1969. The CCD image sensor is at the heart of a digital camera, and it

---

[1] This organization is now known as the National Institute of Standards and Technology (NIST).

Fig. 22.3 First digital
image in 1957

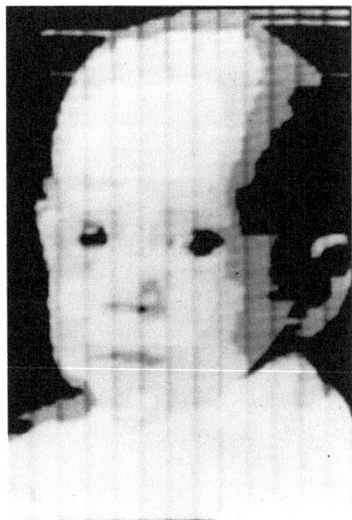

senses light color and intensity and converts it into digital signals. CCDs are widely used today, and while the original CCD had just $100 \times 100 = 10,000$ pixels, modern CCD image sensors have several megapixels.

A *digital camera* is a camera that produces digital images, and most cameras used today are digital. The digital images may be stored on a computer, printed or displayed on a screen. Digital cameras have been incorporated into many electronic devices such as mobile phones, laptops, and tablets.

The first digital camera was invented and built in 1975 by Steven Sasson at Eastman Kodak. It used a charge-coupled device and a lens from a Kodak movie camera, and the black and white image was stored on a cassette tape. The camera weighed 3.6 kg; it could record a 100×100 pixel black and white image (i.e., 0.01 MP); and it took 23 s to record the image on the tape.

Kodak engineers developed a special screen to view the images, and Sasson demonstrated the device to management at Eastman Kodak. They were unimpressed and saw no commercial potential in the proposed device. Sasson and Robin Hills made the first DSLR camera many years later in 1989, but it was not sold, as Kodak was concerned that it would negatively affect its traditional camera business.

A digital camera captures images as pixel elements, where a pixel (i.e., picture element) is the smallest component of a digital image. A megapixel is one million pixels, and the number of colors that can be represented by a pixel depends on the number of bits per pixel. The resolution of an image is dependent on the number of pixels in the camera, with a higher-resolution image requiring more pixels. The quality of a digital photo is influenced by factors such as the optical quality of the lens, the processor and camera sensor, the proper lighting of the subject, the highest resolution on the camera, and the shutter speed.

Bryce Bayer developed a *color filter array* termed the *Bayer filter* at Kodak in the mid-1970s. This allows a single CCD to capture color images and is a way for arranging RGB color filters on a square grid of photo sensors. The particular arrangement of the color filers is used to create a color image. The Bayer color filter eliminates the need to have three separate sensors and represented a paradigm shift in the way of viewing and capturing images. It is the technology behind the digital images captured today, and it played a part in the development of the first digital camera in 1975.

The first true commercial digital camera was the Fujix DS-1P, which was developed by Fuji in Japan and demonstrated at the Photokino Trade Fair in Germany in 1988. It was the first camera to save data to a semiconductor memory card, and its 2MB SRAM memory card could hold 5–10 photographs worth of data. This method of storage was revolutionary for its time but is taken for granted today. The camera had a 400k CCD, which gave it a resolution of 0.4 MP (megapixels). However, the camera was never marketed to the public, and Fuji developed a variant, the Fujix DS-X, which was marketed and sold in Japan.

The JPEG and MPEG standards were defined for digital photography in 1988. The JPEG (Joint Photographic Experts Group) is an information technology digital compression and coding standard for images, and it is the most common format used in digital cameras. The MPEG (Movie Picture Experts Group) is an information technology standard for the coding of moving pictures and audio.

The Kodak Digital Camera System (DCS) is a series of digital single-lens reflex (DSLR) cameras that were based on existing 35 mm film SLRs from Nikon (Fig. 22.4). The original Kodak DCS 100 was the first digital SLR when it was released in 1991, and it had a resolution of 1.3 MP (1320 × 10350 pixels). This professional digital camera system was marketed toward journalists, and it was

Fig. 22.4 Kodak DCS 420
based on Nikon F90 body

initially a very expensive camera and cost over $10,000. The Nikon D1 was a more affordable DSLR camera and was released in 1999.

A plethora of digital cameras with increasing power and features appeared from the 1990s. Companies such as Sony, Olympus, Hitachi, Canon, Nikon, and Kodak introduced them, and there were improvements in image resolution, memory cards, support for JPEG and MPEG standards, and batteries. The price of digital cameras began to fall to acceptable levels and became affordable to consumers in the 1990s.

The Apple QuickTake 100 color digital camera was released in 1994, and it could capture and store eight 640 × 480 pictures. Casio released the QV 10 in 1995, and it was one of the first digital cameras with an LCD display. Kodak introduced the Compact Flash Card in 1994, and the Kodak CD-25 was one of the first cameras to use it for storage when it was introduced in 1996. Hitachi introduced the MPEG1 in 1997, and this was the first digital camera to capture movies in MPEG format.

Camera phones are ubiquitous today, and Samsung developed the first phone with a built-in camera in 2000. It could take 20 photos at a 0.35MP resolution, but it was not possible to send the photos electronically. Further, the phone needed to be connected to the computer to get the photos, as the camera and phone components were essentially separate devices housed in the same body. The first camera phone released in the United States was the Sanyo SCP-5300 in 2002, and this 0.3 MP phone could take pictures with 640 × 480 pixels.

Camera phones took off from 2003 as prices fell to realistic levels, and about 50% of phones had a built-in camera by 2004. Nokia introduced a 2MP camera phone (the N90) in 2005; the Sony Ericsson K800i had a 3.2 MP camera and was released in 2006; and the Nokia N95 had a 5MP camera.

The invention of the smartphone was a revolution in computing, and Apple's original iPhone was released in 2007. It had just a 2MP camera, and the challenge for smartphone makers was to develop a quality camera for a slim smartphone. This has led to several high-end smartphones from Sony, Samsung, and Nokia. For example, the Nokia Lumia 1020 has a 41MP camera.

Camera phones allow users to upload photos to social media sites such as Facebook or Twitter to share events with family or friends or their followers.

Chapter 23
EDVAC and ENIAC Computers

The Electronic Numerical Integrator and Computer (ENIAC) was one of the first general-purpose electronic digital computers, and it was used to integrate ballistic equations and to calculate the trajectories of naval shells (Fig. 23.1). It was completed in 1946 and remained in use until 1955. The original cost of the machine was approximately $500,000.

ENIAC was developed by John Mauchly, Presper Eckert, and others. Mauchly was a lecturer at the Moore School of Electrical Engineering, at the University of Pennsylvania, and Presper Eckert was an engineering student at the school. Mauchly was familiar with Atanasoff's work on the ABC computer (discussed in Chap. 2), and he made a proposal to build an electronic computer using vacuum tubes that would be much faster and more accurate than the existing differential analyzer used in the school (Fig. 1.1). He discussed his idea with representatives with the US Army in 1943, and they agreed to provide the funding to build the machine.

Mauchly focused on the design of the machine and Eckert on the hardware engineering side. ENIAC was one of the first digital computers, and it was a large bulky machine over 100 feet long, 10 feet high, 3 feet deep, and weighed about 30 tons. Its development commenced in 1943 at the University of Pennsylvania, and it was built for the US Army's Ballistics Research Laboratory. It generated a vast quantity of heat, since there were over 18,000 vacuum tubes in the machine, and each vacuum tube generated heat like a light bulb. The machine used 150kW of power, and air conditioning was used to cool it.

It employed decimal numerals, and it could add 5000 numbers and do 357 10-digit multiplications or 35 10-digit divisions in 1 s. It could be programmed to perform complex sequences of operations, such as loops, branches, and subroutines. However, the task of taking a problem and mapping it onto the machine was complex, and it usually took several weeks to perform. The first step was to determine

© Springer Nature Switzerland AG 2018
G. O'Regan, *The Innovation in Computing Companion*,
https://doi.org/10.1007/978-3-030-02619-6_23

Fig 23.1 Setting the switches on ENIAC's function tables. US Army photo

what the program was to do on paper; the second step was the process of manipulating the switches and cables to enter the program into ENIAC, which took several days. The final step was verification and debugging, and this involved single-step execution of the machine.

Mauchly and Eckert were aware of the limitations of ENIAC, and they began work on the design of a successor computer, EDVAC, in early 1945. ENIAC had to be physically rewired to perform different tasks, and there was a need for an architecture that would allow a machine to perform different tasks without having to go through a time-consuming rewiring process. This led to the concept of the *stored program*, which was implemented on EDVAC. The idea of a stored program is that the program is stored in memory, and when there is a need to change the task to be performed, then all that is required is to place a new program in memory rather than rewiring the machine.

Von Neumann became involved in some of the engineering discussions, during the development of EDVAC, and he produced a draft report describing this computer. The report was intended to be internal, but circumstances changed due to legal issues that arose with respect to the ownership of intellectual property and patents. This led to the resignation of Mauchly and Eckert from the Moore School of Electrical Engineering, as they wished to protect their patents on ENIAC and EDVAC. They set up their own company (EMCC) to exploit the new computer technology (see Chap. 19).

The Moore School removed the names of Mauchly and Eckert from the internal report, and circulated it to the wider community. It mentioned the fundamental computer architecture that is known as the *von Neumann architecture* (see Chap. 54), and Mauchly and Eckert received no acknowledgement for their contributions. The concept of a stored program and von Neumann architecture is detailed in von Neumann (1945).

EDVAC (the successor of ENIAC) implemented the concept of a stored program in 1949, just after its implementation on the Manchester Baby prototype machine in England.

There were problems initially with the reliability of ENIAC as several vacuum tubes burned out most days (Fig. 23.2). This meant that the machine was often non-functional, as high-reliability tubes did not become available until the late 1940s. However, most of the problems occurred during the warm-up and cool-down periods, and it was decided not to turn the machine off. This led to improvements in its reliability to the acceptable level of one tube every 2 days. The longest continuous period of operation without a failure was 5 days.

The very first program run on ENIAC took just 20 s, and the answer was manually verified to be correct after 40 h of work with a mechanical calculator. One of the earliest problems solved was related to the feasibility of the hydrogen bomb. It involved the input of 500,000 punch cards, and the program ran for 6 weeks and gave an affirmative reply. The Atanasoff-Berry Computer (ABC), the Colossus computer in the United Kingdom, and the Z3 computer in Germany all preceded ENIAC in development. ENIAC is a major milestone in the history of computing.

Fig 23.2 Replacing a valve on ENIAC. US Army photo

Fig. 23.3 The EDVAC computer. US Army photo

23.1 EDVAC Computer

EDVAC (Electronic Discrete Variable Automatic Computer) was the successor to the ENIAC computer (Fig. 23.3). It was a stored program computer and it cost $500,000. Eckert and Mauchly proposed it in 1944, and design work commenced prior to the completion of ENIAC.

It was delivered to the Ballistics Research Laboratory in 1949 and commenced operations in 1951. It employed 6000 vacuum tubes, had a power consumption of 56,000 watts, and had 5.5 Kb of memory. EDVAC was one of the earliest stored-program computers, with the program instructions held in memory, and the machine did not need to be physically rewired to solve a different problem. EDVAC remained in operation until 1961.

23.2 Controversy ABC and ENIAC

Mauchly became embroiled in a legal dispute in the 1973 *Honeywell vs. Sperry Rand* patent court case in the United States. This controversy arose from a patent dispute between Sperry and Honeywell, and John Atanasoff (discussed in Chap. 2) was called as an expert witness in the case.

The ABC computer was ruled to be the first electronic digital computer in the 1973 judgment. The court ruled that Eckert and Mauchly did not invent the first electronic computer, since the ABC Computer existed as *prior art* at the time of their patent application. It is fundamental in patent law that an invention is novel and that there is no existing prior art. This meant that Mauchly and Eckert's patent application for ENIAC was invalid, and John Atanasoff was named as the inventor of the first digital computer.

Mauchly had visited Atanasoff and Berry on several occasions prior to the development of ENIAC, and they had discussed the implementation of the ABC computer. The court ruled that the ABC was the first digital computer, and that the inventors of ENIAC had derived the subject matter of the electronic digital computer from Atanasoff.

Chapter 24
Eliza Program

Joseph Weizenbaum (Fig. 24.1) developed the Eliza program at MIT in 1966, which was one of the earliest AI programs. This famous natural language understanding program is an important milestone in the AI field, and it was named after the character *Eliza*. in the musical *My Fair Lady*, which was based on Shaw's 1912 play *Pygmalion*.

The Eliza program simulated a conversation between a patient and a psychotherapist, with the program using the person's response to shape its reply. The interaction was between the computer program and a user sitting at an electric typewriter, with the user typing and the computer program responding. The program convinced several users that it had real understanding and that it was an empathic psychotherapist. This led to users unburdening themselves in long computer sessions. Eliza was a sensation at the MIT campus, and it rapidly spread to other universities.

Weizenbaum was a German-American computer scientist who developed the Eliza program, and he became well known for his views on the ethics of artificial intelligence. He became a leading critic of the AI field following the response of users to the Eliza program. Many users felt that they were communicating with an empathic psychologist rather than a machine, and Weizenbaum was shocked to discover that so many users were taking his program seriously and that they were sharing their most private thoughts with the machine. He became an advocate of social responsibility in science and a leading critic of AI research.

Weizenbaum's Eliza program demonstrated the threat that AI poses to privacy. It is conceivable that an AI program may be developed in the future that is capable of understanding speech and natural languages. Such a program could theoretically eavesdrop on every phone conversation and email and gather private information on what is said and who is saying it. Such a program could be used by a state to suppress dissent and to eliminate those who pose a threat to its existence.

The Eliza program had no understanding, and it used pattern matching to take the user's input and to send output that would continue the conversation. Weizenbaum had joined General Electric in 1955, and he developed a list processing system called SLIP (Symmetric List Processor) in 1963. He became professor of computer

© Springer Nature Switzerland AG 2018

G. O'Regan, *The Innovation in Computing Companion*,
https://doi.org/10.1007/978-3-030-02619-6_24

Fig 24.1 Joseph
Weizenbaum

science at MIT in 1963 and used his knowledge of SLIP to write the Eliza program
at MIT in 1966.

He wrote a critique of AI in his influential book *Computer Power and Human
Reason* Weizenbaum (1976). This popular book deals with the role of computers in
society. He argues that while artificial intelligence may someday be possible, computers should never be given responsibility for making important human decisions,
as they lack human qualities such as compassion and wisdom. The book raised
questions on the role of artificial intelligence and led to debate on the role of computer systems in decision-making and on the reliance on computer systems and
technology in the world today.

24.1   Eliza

Weizenbaum created the famous Eliza program at MIT in 1966, and the program
interacted with a user sitting at an electric typewriter, in English, in the manner of a
Rogerian psychotherapist.[1]

The name *Eliza*[2] was chosen because it could be taught to speak increasingly
well, as in Shaw's Pygmalion. It was one of the earliest natural language processing
programs, and Weizenbaum gave an example of a conversation between Eliza and a

---

[1] Rogerian psychotherapy is a form of psychotherapy developed by Carl Rodgers in the 1940s and
1950s. It is also known as person-centered therapy.

[2] Eliza Doolittle was a working-class character in Shaw's play, Pygmalion. She is taught to speak
with an upper-class English accent by Professor Henry Higgins.

young woman in Weizenbaum (1966). Many users believed that Eliza had real understanding, and they began to unburden themselves in long computer sessions.

The program operated by decomposing the user's input into its constituent parts of speech and then typing them back at the user in a manner to continue with the dialogue. The program provided very convincing human-like interaction, as may be seen in the sample dialogue below:

User/Patient:	Men are all alike.
Eliza	*In what way?*
User/Patient:	They are always bugging us about something or other.
Eliza	*Can you think of a specific example?*
User/Patient:	Well, my boyfriend made me come here.
Eliza	*Your boyfriend made you come here?*
User/Patient	He says I'm depressed much of the time.
Eliza	*I am sorry to hear that you are depressed.*
User/Patient	It's true, I am unhappy.
Eliza	*Do you think that coming here will help you not to be unhappy ?*
User/Patient	I need some help, that much seems certain.
Eliza	*What would it mean to you if you got some help?*
User/Patient	Perhaps, I could learn to get along with my mother.
Eliza	*Tell me more about your family.*

Weizenbaum was shocked to discover that so many users were convinced that the program had real understanding, and users spent hours sharing their personal problems with the program. It led Weizenbaum to think about the ethics and implications of the artificial intelligence field and the consequences of a relatively trivial program deceiving a naïve user to reveal personal information. He became a leading critic of the AI field and an advocate for professional and social responsibility.

Even though the conversation above gives the impression that the program had real understanding, the following parody of Eliza in another dialogue shows the level of understanding of the program:

Eliza	*How do you do. Please state your problem.*
User/Patient:	I'm wondering whether computers can think.
Eliza	*Why do you mention computers?*
User/Patient:	Since you are one I thought I would ask you - can you think?
Eliza	*Did you think they might not be one you thought you would ask – can I think*
User/Patient:	Let's talk just about you - not me. Can you think?
Eliza	*You like to think I – Not You – Don't You*
User/Patient:	Pardon?
Eliza	*What does it suggest to you?*
User/Patient:	What does it suggest to you?
Eliza	*Tell me about what does it suggest to I?*
User/Patient:	That's nonsense – which of us do you mean by I?
Eliza	*Why do you think I mean by you*
User/Patient:	The question meant to which of us is the word referring.

Eliza	*Suppose the question meant to which of us were not the word referring*
User/Patient:	I think you are an extremely stupid program.
Eliza	Perhaps you would like to be an extremely stupid program

24.2   Eliza and the Ethics of AI

Weizenbaum's views on the ethics of AI are discussed in his book *Computer Power and Human Reason* Weizenbaum (1976). He displays ambivalence toward computer technology and argues that it is dangerous and immoral to assume that computers could eventually take over any role, given sufficient processing power and appropriate programming.

He argues that AI is a threat to human dignity and that AI should not replace humans in positions that require respect and care. He states that machines lack empathy and that if they replace humans in positions such as police officers or judges, this would lead to alienation and a devaluation of the human condition. Weizenbaum's views are controversial and have been criticized by John McCarthy and others.

Weizenbaum's Eliza program demonstrated the threat that AI poses to privacy. It is conceivable that an AI program may be developed in the future that is capable of understanding speech and natural languages. Such a program could theoretically eavesdrop on and gather private information on what is said and who is saying it.

As more and more sophisticated machines and robots are created, it is, of course, essential that intelligent machines behave ethically and have a moral compass to distinguish right from wrong. It remains an open question as to how to teach a robot right from wrong, and the science fiction writer, Asimov, proposed three laws of robotics in his book, *I Robot*.

These laws basically state that a robot must not injure a human being either through action or inaction. A robot is required to obey orders (provided that the order does not involve harm to another human being), and finally a robot must protect its own existence unless of course its continued existence will cause harm to humans.

Chapter 25
Email Communication

The transfer of information between people and societies goes back to early civilizations. The Egyptian pharaohs used couriers to communicate new laws throughout the state from around 2500 BC. The Persian Empire established an early postal system in the sixth century B.C., and the Chinese established an early postal system from the third century B.C. The Romans established a postal system (initially just used by the government) from the first century B.C.

A pigeon messaging system was used to send messages faster from the twelfth century, where the homing characteristic of pigeons was the underlying principle employed for transmission. Another approach adopted was the semaphore system (from the fourth century B.C.), which allowed very simple messages to be exchanged between groups on two different hills (similar in a sense to smoke signals). A ship semaphore system was introduced in the fifteenth century, which allowed two ships to communicate with each other (this system used flags where the position and motion of a flag represented a letter).

The Chappe brothers introduced an early optical telegraph system in Europe in the late eighteenth century, and early electrical telegraph systems were introduced in the early nineteenth century. Samuel Morse devised a system (Morse code) that allowed letters to be represented by a series of on-off tones in the late 1830s. Marconi introduced a system for the wireless transmission of sounds in 1896, and this led to the beginning of communication between ships at sea and coastal radio stations.

The telephone was invented by Alexander Graham Bell in 1876, and its invention was a paradigm shift from that of letter writing or face-to-face meetings to the new paradigm of instantaneous communication with another person. The telex network (a global public switched network of teleprinters) was introduced in the late 1920s, and it was used by businesses for sending text-based messages. Telex remained in use up to the late twentieth century until the fax machine led to its decline and eventual demise.

Postal systems were introduced into Europe and America from the nineteenth century. The early postal system in the United Kingdom required the recipient

© Springer Nature Switzerland AG 2018
G. O'Regan, *The Innovation in Computing Companion*,
https://doi.org/10.1007/978-3-030-02619-6_25

(rather than the sender) to pay for the letter, and the charge applied depended on the number of pages and the distance traveled. The system was reformed in 1840 with the introduction of the prepayment and penny postage.

Electronic mail (or email) is a method used by the sender to exchange electronic messages, using digital devices such as computers, laptops, tablets, and mobile phones to prepare and send the message to the recipient. The origin of electronic mail goes back to ARPANET, which was developed, in the late 1960s, and when standards for encoding email messages were published.

25.1   Invention of Email

There were approximately 10,000 computers in the world in the 1960s, and communication between them was virtually nonexistent. However, several computer scientists had dreams of worldwide networks of computers, where every computer around the globe would be interconnected to all others. Licklider[1] wrote memos on his concept of an intergalactic network in the early 1960s, which envisaged that everyone around the globe would be interconnected and able to access programs and data at any site from anywhere Licklider (1960).

The US Department of Defense founded the Advanced Research Projects Agency (ARPA) in the late 1950s. Licklider became the head of its computer research program, and he developed close links with MIT, UCLA, and BBN Technologies.[2] The concept of packet switching[3] was invented in the 1960s, and several organizations commenced work on its implementation.

The first (nonmilitary) wide-area network connection was created (using a dedicated telephone wire for data transfer) in 1965,[4] with a connection between a computer at MIT and a computer in Santa Monica. ARPA recognized the need to build a network of computers in the mid-1960s, and this led to the ARPANET project (the precursor to the Internet), which aimed to implement a packet-switched network with a network speed of 56 Kbps. ARPANET became the world's first packet-switched network when it was introduced in the late 1960s, and it remained operational until 1990 (when the Internet became took off).

BBN Technologies implemented the network, and interconnected *Interface Message Processor* (IMPs) performed the network management. These were in the

---

[1] Licklider was an early pioneer of AI and wrote an influential paper "Man-Computer Symbiosis" in 1960, which outlined the need for simple interaction between users and computers.

[2] BBN Technologies (originally Bolt Beranek and Newman) is a research and development technology company. It played an important role in the development of packet switching and in the implementation and operation of ARPANET. The "@" sign used in an email address was a BBN innovation. BBN became a subsidiary of Raytheon in 2009.

[3] Packet switching is a message communication system between computers. Long messages are split into packets, which are then sent separately to minimize the risk of congestion.

[4] The SAGE system did early work done on wide-area networks in the late 1950s.

front of the main computers, and they evolved over time to become the network routers that are used today.

The *Network Working Group* at UCLA developed a set of rules that specified how the computers on the network should communicate. These rules were called the *Network Control Protocol* (NCP). The first host-to-host connection was made between a computer in UCLA and a computer at SRI in late 1969. Several other nodes were added to the network until it reached its target of 19 nodes in 1971.

The Network Working Group developed the *telnet protocol* and the *file transfer protocol* (FTP) in 1971. A public demonstration of ARPANET was made in 1972, and one of the earliest demos was that of Weizenbaum's ELIZA program (see Chap. 24).

Ray Tomlinson (Fig. 25.1) of BBN Technologies is recognized as the inventor of modern email, as he developed an email program at BBN that allowed a user to send electronic mail to another user who was connected to a different host machine on ARPANET. Tomlinson sent the first text letter between two ARPANET-connected computers in 1971.

The users of existing email systems could only send messages to others that used the same mainframe computer, and Tomlinson introduced the @ sign to specify the machine name of the user where the message should be sent to (e.g., bob@mit). The addresses were initially of the form username@hostname, but this was later revised to username@host.domain with the development of the domain name system (DNS).

Fig 25.1 Ray Tomlinson. Creative Commons

Tomlinson's email system was a major advance on existing email systems used in organizations, and it led to a revolution in the way that people communicate. Email systems are based on a store-and-forward model where the email server accepts, stores, forwards, and delivers messages, and neither the users nor their computers are required to be online. There are several protocols used with email systems including *Simple Mail Transfer Protocol* (SMTP), the *Post Office Protocol* (POP3), and the *Internet Message Application Protocol* (IMAP).

Messages are exchanged between hosts using the SMTP protocol, and the destination address consists of the username and the domain name. The system then attempts to deliver the message, and where it cannot be delivered the message bounces back to the sender indicating a problem. Users may retrieve their messages from the server using POP or IMAP protocols.

It is important to use email appropriately, and care is often required before replying to an email, especially as it easy to appear as abrupt or harsh in correspondence. Further, emails leave a trail and may be forwarded by the recipient, and so it is essential that whatever is written does not cause injury to others. Therefore, an email should be courteous and should be written professionally in the workplace. For example, it should include a subject line and address the person (or audience) appropriately. It needs to be sensitive to cultural differences, and care may be required with humor. Finally, the spelling should be checked, and it should be reread prior to it being sent to ensure that the right tone is set in the communication.

25.2   Gmail

Google Mail is the most widely use web-based email service (Yahoo, Hotmail, and Outlook are other web-based email systems), with over a billion users around the world. The user logs into the web-based email account using a web browser to send and receive mail. Google provides over 15 GB of free storage between Gmail, Google drive, and Google+, and users can purchase additional storage as required up to a maximum of 300 TB. Gmail is available on personal computers as well as on tablets and mobile devices, and over 50 languages are supported.

Gmail includes a search bar for searching emails, and it also allows web searches to be performed. It automatically scans all incoming and outgoing mail for viruses in email attachments, and it will prevent the message from being sent if a virus is found in the outgoing attachment. Further, it will attempt to remove any viruses found in an incoming email attachment.

Gmail automatically scans the contents of emails to add context-sensitive advertisements to them and to filter spam. This has raised privacy concerns as it means that all emails sent or received are scanned and read by some computer, and Google has stated in court filings that no *reasonable expectation exists among Gmail users with respect to the confidentiality of their emails*. Further, Google argues that the automated scanning of emails is done for the benefits of the user, as it allows Google to provide customized search results, tailored advertisements, and the prevention of spam and viruses.

Chapter 26
E-Commerce

One of the most popular activities on the Web is shopping, which allows customers the flexibility to purchase products at their leisure, as well as allowing merchants the opportunity of creating a large database of attractive products. The process of buying or selling products online is termed e-commerce, and these transactions are carried out over the World Wide Web.

Michael Aldrich described the first online shopping system in 1979. He coined the term *teleshopping* (shopping at a distance), when he connected a domestic television to a real-time transaction-processing computer in the late 1970s. The inspiration for his idea was a conversation he had with his wife on finding a simpler way to do the supermarket shopping. Today, teleshopping is ubiquitous, and it is now known as online shopping.

One of the earliest pre-World Wide Web online services was the Minitel system, which was introduced in France in the early 1980s. France Telecom designed this popular system, and it allowed users to make online purchases, make train reservations, check stock prices, and have a mail box. France Telecom distributed Minitel terminals to telephone subscribers, and this enabled users to access a nationwide electronic network of telephone and address information. Minitel was introduced into several other countries in the late 1980s/early 1990s.

Tim Berners-Lee invented the World Wide Web in the late 1980s, and he realized that the Web offered the potential to conduct business in cyberspace, rather than the traditional way where buyers and sellers do business at the marketplace. The success of the Web led to modern electronic commerce, and the required technologies (e.g., electronic funds transfer, secure socket layer, electronic data interchange, online transaction processing) were later developed.

The growth of the Web was phenomenal, and exponential growth rate curves became a feature of newly formed Internet companies and their business plans. It was predicted that the new web-based economy would replace traditional bricks and mortar companies, and it was expected that most business would be conducted over the web, with traditional enterprises losing market share and going out of business.

© Springer Nature Switzerland AG 2018
G. O'Regan, *The Innovation in Computing Companion*,
https://doi.org/10.1007/978-3-030-02619-6_26

Table 26.1 Characteristics of e-commerce

Feature	Description
Catalogue of products	The catalogue of products details the products available for sale and their prices
Well-designed and easy to use	This is essential as otherwise the web site will not be used
Shopping carts	This is analogous to shopping carts in a supermarket
Security	Security of credit card information is a key concern for users of the web, as users need to have confidence that their credit card details will remain secure.
Payments	Once the user has completed the selection of purchases, there is a checkout facility to arrange for the purchase of the goods
Order Fulfillment/ order inquiry	Once payment has been received the products must be delivered to the customer

Exponential growth of e-commerce companies was predicted, and the size of the new web economy was estimated to be in trillions of US dollars.

New companies were formed to exploit the commercial opportunities of the Web, and existing companies developed e-business and e-commerce strategies to adapt to the brave new world. Companies providing full e-commerce solutions began selling of products or services over the web to either businesses or consumers. These business models are referred to as business-to-business (B2B) or business-to-consumer (B2C). E-commerce web sites have the following characteristics (Table 26.1).

26.1   Formation of Dot Com Companies

The success of the Web was phenomenal, and it led to a boom in the formation of *new economy* businesses. These included the Internet portal company, Yahoo; the online book store, Amazon; and the online auction site, eBay. Yahoo provides news and a range of services, and most of its revenue comes from advertisements. Amazon initially sold books, but it now sells a collection of consumer and electronic goods. eBay brings buyers and sellers together in an online auction space.

Some of these new technology companies were successful and remain in business. Others were financial disasters due to poor business models, poor management, and poor implementation of the new technology. Some of these companies offered an Internet version of a traditional bricks and mortar company, whereas others had a unique business offering. For example, eBay offers an online auctioneering site to customers worldwide, which was a totally new service and quite distinct from traditional auctioneering.

Jeff Bezos founded Amazon in 1995 as an online bookstore. Its product portfolio has expanded to include just about anything (e.g., CDs, DVDs, toys, computer software, video games, and a cloud computing platform). Its initial focus was to build up the "Amazon" brand throughout the world and to become the world's largest bookstore. It initially sold books at a loss by giving discounts to buyers to build up market share. It was very effective in building its brand through advertisements, marketing, and discounts. It has a solid business model with a very large product catalogue, a well-designed web site with good searching and check out facilities, and good order fulfillment.

Pierre Omidyar founded eBay in 1995, and the site brings buyers and sellers together. Millions of items are listed, bought, and sold on eBay every day. The sellers are individuals or international companies. Any legal product that does not violate the company's terms of service may be bought or sold on the site. A buyer makes a bid for a product or service and competes against several other bidders. The highest bid is successful, and payment and delivery are then arranged. The revenue earned by eBay includes fees to list a product and commission fees that are applied whenever a product is sold.

26.1.1   Dot Com Failures

Several of the companies formed during the dot com era were successful and remain in business. Others had inappropriate business models or poor management and failed in a spectacular fashion.

Webvan.com was an online grocery business based in California. It delivered products to a customer's home within a 30-minute period of their choosing. The company expanded to several other cities before it went bankrupt in 2001. Many of its failings were due to management as the business model was reasonable, and today there are several successful online fresh food delivery businesses. The management was inexperienced in the supermarket or grocery business, and the company spent excessively on infrastructure. It had been advised to build up an infrastructure to deliver groceries as quickly as possible, rather than developing partnerships with existing supermarkets. It built warehouses, purchased a fleet of delivery vehicles and top of the range computer infrastructure before running out of money.

Ernst Malmsten and others founded Boo.com in 1998, as an online fashion retailer based in the United Kingdom. Its web site was poorly designed for its target audience, and it went against many of the accepted usability conventions of the time. The web site was designed in the days before broadband, with 56K modems used by most customers. However, its design included the latest Java and Flash technologies, and it took most users several minutes to load the first page of the web site. Further, the navigation of the web site was inconsistent, and it changed as the user moved around the site.

Other reasons for failure included poor management and leadership, lack of direction, lack of communication between departments, spiraling costs left unchecked, and crippling payroll costs. Further, purchasers returned many products, and there was no postage charge applied for this service. The company spent over

$135 million of shareholder funds in less than 3 years, and it went bankrupt in 2000. There is an account of its formation and collapse is in the book, *Boo Hoo* (Malmsten and Portanger 2002). This book is a software development horror story, and it shows the company's immaturity in developing software for a state-of-the-art web site. The net effect was that despite extensive advertising by the company, the web site was not fit for purpose and users were not inclined to use it.

Pets.com was an online pet supply company founded in 1998 by Greg McLemore. It sold pet accessories and supplies. It had a well-known advertisement as to "Why one should shop at an online pet store?". The answer to this question was: "Because Pets Can't Drive!". Its mascot (the Pets.com sock puppet) was well known. It launched its IPO in February 2000 just before the dot com collapse.

Pets.com made investments in infrastructure such as warehousing and vehicles. It needed a critical mass of customers to break even, and its management believed that it needed $300 million of revenue to achieve this. They expected that this would take a minimum of 4–5 years, and therefore there was a need to raise further capital. However, following the dot com collapse, there was negative sentiment toward technology companies, and it was apparent that it would be unable to raise further capital. The company went into liquidation 9 months after its IPO.

Joseph Park and Yong Kang founded Kozmo.com in New York in 1998. It was an online company that promised free 1-hour delivery of small consumer goods. It provided point-to-point delivery usually on a bicycle and did not charge a delivery fee. Its business model was deeply flawed, as it is expensive to offer point-to-point delivery of small goods within a 1-hour period without charging a delivery fee. The company argued that they could make savings to offset the delivery costs, as they did not require retail space. It raised about $280 million from investors and ceased trading in 2001.

26.1.2 Bubble and Burst

The initial public offering (IPO) of Netscape in 1995 demonstrated the incredible value of the new Internet companies. Netscape had planned to issue the share price at $14, but it decided at the last minute to issue it at $28. The share price reached $75 later that day. This was followed by what became the dot com bubble with many public offerings of Internet stock and where the value of these stocks reached astronomical levels. Reality returned to the stock market when it crashed in April 2000, and share values returned to more realistic levels.

Most of these Internet companies were losing substantial sums of money, and few expected to deliver profits in the short term. Financial instruments such as the balance sheet, profit and loss account, and price to earnings ratio are normally employed in estimating the value of a company. However, investment bankers argued that there was a new paradigm in stock market valuation for Internet companies, which suggested that the potential future earnings of technology companies be considered in determining their value. This was used to justify the high prices of shares, as frenzied investors rushed to buy these overpriced and overhyped stocks. Common sense seemed to play no role in decision-making, and the bubble was characterized by:

- Irrational exuberance on the part of investors
- Insatiable appetite for Internet stocks
- Incredible greed from all parties involved
- Following herd mentality
- A lack of rationality and common sense by all concerned
- Traditional method of company valuation not employed
- Interest in making money rather than in building the business first
- Questionable decisions by Federal Reserve Chairman (Alan Greenspan)
- Questionable analysis by investment firms
- Conflict of interest (investment banks)

There were winners and losers in the boom and collapse. Some investors made a lot of money from the bubble, with others including pension funds and life assurance funds making significant losses. The investment banks typically earned 5–7% commission on each successful IPO, and it was therefore not in their interest to question the boom too closely. Those who bought and sold early made a good return, whereas those who kept their shares for too long suffered major losses. The extent of the boom can be seen in the rise and fall of the value of the Dow Jones and NASDAQ from 1995 through 2002.

The extraordinary rise of the Dow Jones (Fig. 26.1) represented a 200% increase over 5 years. The rise of the NASDAQ (Fig.26.2) over this period is even more dramatic, representing a 566% increase during the period.

The fall of the indices was equally as dramatic especially in the case of the NASDAQ. It was clear that Internet companies were rapidly going through the cash raised at the IPOs and that a significant number would be out of cash by the end of 2000. Therefore, these companies would either go out of business or would need to go back to the market for further funding. This led to questioning of the relatively

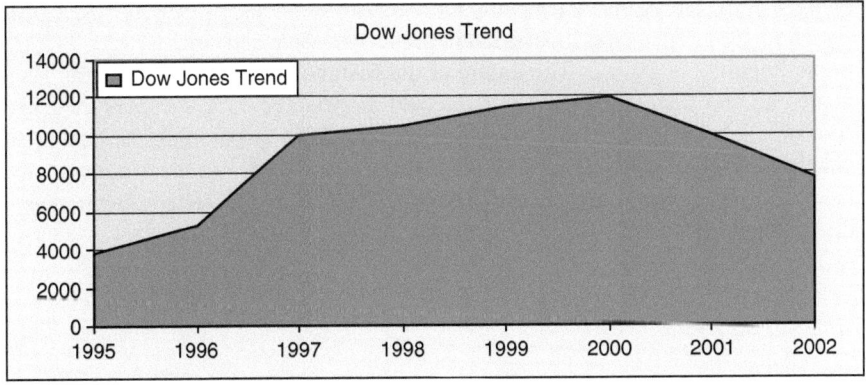

Fig. 26.1 Dow Jones (1995–2002)

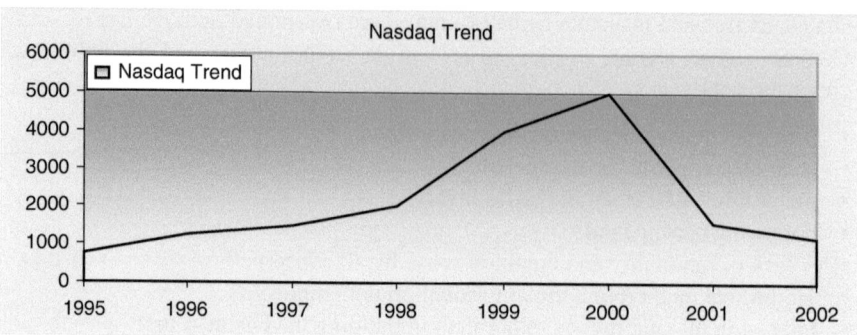

Fig. 26.2 NASDAQ (1995–2002)

unquestioned business models of these firms. Funding is easy to obtain when stock prices are rising at a rapid rate. However, when prices are static or falling, with negligible or negative business return to the investor, then funding dries up. The actions of the Federal Reserve in rising interest rates to prevent inflationary pressures also helped to correct the market.

Some independent commentators had recognized the bubble, but their comments and analysis had been largely ignored. These included *The Financial Times* and the *Economist*, as well as some commentators in the investment banks. Investors rarely queried the upbeat analysis coming from Wall Street and seemed to believe that rising stock prices would be a permanent feature of the US stock markets. Greenspan had argued that it is difficult to predict a bubble until after the event and that even if the bubble had been identified, it could not have been corrected without causing a contraction. Instead, the responsibility of the Fed according to Greenspan was to mitigate the fallout when it occurs.

There have, of course, been other stock market bubbles throughout history. For example, in the 1800s, there was a rush on railway stock in England leading to a bubble and eventual burst of railway stock prices in the 1840s. There was the famous Tulip Mania bubble in the Netherlands in the seventeenth century, where the price of bulbs of the newly introduced tulip reached astronomical levels before prices collapsed in 1637. There was a devastating property bubble and collapse (2002–2009) in the Republic of Ireland. The failure of the Irish political class, the Central Bank of Ireland and Irish financial regulators, and the Irish banking sector in their irresponsible lending policies and failures of the media in questioning the bubble are deeply disturbing. Its legacy remains, and while the country has made a remarkable recovery, the failures of so many at senior level in the state remain deeply disturbing.

Chapter 27
Formal Methods

Formal methods are an innovative technology used in the development of safety critical software. The term *formal methods* refer to various mathematical techniques used for the formal specification and development of software. They consist of a formal specification language, and a collection of tools that support the syntax checking of the specification, as well as the proof of properties of the specification. Formal methods allow questions to be asked about what the system does independently of its implementation.

The use of mathematical notation helps to avoid the problem of ambiguity inherent in a natural language description, since mathematics is precise and rigorous. The formal specification becomes the key reference point for the different parties involved in the construction of the system: e.g., for those involved in the requirements, program implementation, and testing and program documentation. It thus promotes a common understanding of the system for all those involved.

The term *formal methods* is used to describe a formal specification language and a method for the design and implementation of a computer system. They may be employed at several levels including:

- Formal specification only (program developed informally)
- Formal specification, refinement, and verification (some proofs)
- Formal specification, refinement, and verification (theorem proving)

The specification is written in a mathematical language, and the implementation may be derived from the specification via stepwise refinement. The refinement step makes the specification closer to the implementation, and there is an associated proof obligation to demonstrate that the refinement is valid and that the concrete state preserves the properties of the abstract state. Thus, if the original specification is correct and the proof of correctness of each refinement step is valid, then there is a very high degree of confidence in the correctness of the implemented software.

© Springer Nature Switzerland AG 2018
G. O'Regan, *The Innovation in Computing Companion*,
https://doi.org/10.1007/978-3-030-02619-6_27

Stepwise refinement is illustrated as follows: the initial specification S is the initial model M_0; it is then refined into the more concrete model M_1, and M_1 is then refined into M_2 and so on until the eventual implementation $M_n = E$ is produced.

$$S = M_0 \sqsubseteq M_1 \sqsubseteq M_2 \sqsubseteq M_3 \sqsubseteq \ldots\ldots . \quad \sqsubseteq M_n = E$$

Requirements are the foundation of the system to be built, and the product will be incorrect if the requirements are incorrect. The objective of requirements validation is to ensure that the requirements are those desired by the customer (i.e., to build the right system). Formal methods may be employed to model the requirements, and the model exploration yields further desirable or undesirable properties.

Formal methods may be employed in requirements validation, and they may, in a sense, be used to debug the requirements. They may be used to prove that certain properties are true of the formal specification, and this is important in the safety critical field. These properties may be proved mathematically, and the mathematical analysis may lead to amendments to the requirements.

The use of formal methods generally leads to increased confidence in the correctness of the software. They may be employed at different levels (e.g., just for specification with the program developed informally). However, there are challenges to the deployment of formal methods, as the use of these mathematical techniques may be a culture shock to many employees (depending on the sector in which they are employed).

Formal methods have been applied to a diverse range of applications, including the safety and security critical fields to develop dependable software. The applications include the railway sector, microprocessor verification, the specification of standards, and the specification and verification of programs. Parnas has criticized formal methods on the following grounds (Table 27.1).

However, the use of a formal method such as Z or VDM forces the software engineer to be precise and helps to avoid ambiguities present in natural language. Clearly, a formal specification should be subject to peer review to provide confidence in its correctness. New formalisms need to be intuitive and usable, and an advantage of classical mathematics is that it is familiar to students.

Table 27.1 Criticisms of formal methods

No.	Criticism
1.	Often the formal specification is as difficult to read as the program
2.	Many formal specifications are wrong
3.	Formal methods are strong on syntax but provide little assistance on what technical information should be recorded using the syntax
4.	Formal specifications provide a model of the proposed system. However, a precise statement of the requirements is what is needed
5.	Stepwise refinement is unrealistic. It is like deriving a bridge from a description of a river and the expected traffic on the bridge
6.	Many unnecessary mathematical formalisms have been developed rather than using the available classical mathematics

There is a strong motivation to use best practice in software engineering, and formal methods are a leading-edge technology that may benefit companies. Brown (1990) argues that for the safety critical field that:

> Missile systems must be presumed dangerous until shown to be safe, and that the absence of evidence for the existence of dangerous errors does not amount to evidence for the absence of danger.

This suggests that companies in the safety critical field will need to demonstrate that every reasonable step was taken to prevent the occurrence of defects. One such best practice is the use of formal methods, and its exclusion may need to be justified in some domains. It is possible for a software company to be sued for defective software which injures a third party, and therefore companies need to have a rigorous software quality assurance system in place to detect and prevent defects.

Formal methods have been employed to verify correctness in several domains such as the safety and security critical fields. This includes applications in the nuclear power industry, the aerospace industry, the security technology area, and the railroad domain. These sectors are subject to stringent regulatory controls to ensure that safety and security are properly addressed.

IBM developed the VDM specification language at its laboratory in Vienna, and it piloted the Z formal specification language on the CICS (Customer Information Control System) project at its plant in Hursley, England. There is some evidence to suggest that the use of formal methods reduces costs of the software development. For example, a 9% cost saving is attributed to the use of formal methods during the CICS project (Hinchey and Bowen 1995).

The mathematical techniques developed by Parnas (e.g., his requirements model and tabular expressions) have been employed to specify the requirements of the A-7 aircraft as part of a research project for the US Navy. Tabular expressions were also employed for the software inspection of the automated shutdown software of the Darlington Nuclear power plant in Canada. These are two successful uses of mathematical techniques in software engineering.

Formal methods have been applied in the railway domain, and examples of modeling and verification of a railroad gate controller and railway signaling are described in Hinchey and Bowen (1995). Clearly, it is essential to verify safety critical properties such as *when the train goes through the level crossing then the gate is closed*.

The debate concerning the level of use of mathematics in software engineering is ongoing. Many practitioners are against their use and employ methodologies such as software inspections and testing, or lightweight approaches such as Agile to improve confidence in the correctness of the software. They argue that in the current competitive industrial environment, where time to market is a key driver, the use of such formal mathematical techniques would seriously impact the market opportunity. Industrialists often need to balance conflicting needs such as quality, cost, and on time delivery. They argue that commercial pressures require methodologies that allow them to achieve their business goals effectively.

The other camp argues that the use of mathematics is essential in the delivery of high-quality and reliable software and that if a company does not place sufficient emphasis on quality, it will pay the price in terms of poor quality and its reputation in the marketplace.

It is accepted that formal methods must play a role in the safety critical and security critical fields. Apart from that the extent of the use of mathematics is a hotly disputed topic. Companies face significant competitive forces in a global marketplace, and they need clear evidence that formal methods will support them in delivering products to the marketplace ahead of their competition and at the right price and with the right quality. Formal methods need to prove that it can do this if it wishes to be taken seriously in mainstream software engineering.

There are practical difficulties associated with the industrial use of formal methods. It is usually possible to get a developer to learn a formal method, as a programmer has some experience of mathematics and logic. However, it is often more difficult to get a customer to learn a formal method, and this makes it difficult to perform a rigorous validation of the formal specification.

This means that often a formal specification of the requirements and an informal definition of the requirements using a natural language are maintained. It is essential that both of these are consistent and that there is a rigorous validation of the formal specification. Otherwise, if the programmer proves the correctness of the code with respect to the formal specification, and the formal specification is incorrect, then the formal development of the software is incorrect. There are several techniques to validate a formal specification:

- Proof that the formal specification satisfies key properties
- Software inspections of formal specification/informal set of requirements
- Specification animation to validate the formal specification

Formal methods are perceived as being difficult to use and of providing limited value in mainstream software engineering. Programmers receive education in mathematics, but many never use formal methods again in an industrial position. Some of the reasons for this are:

- Notation is not intuitive.
- It is difficult to write a formal specification.
- Validation of a formal specification is difficult.
- Refinement and proof are difficult.
- Limited tool support.

It is important to design more usable notations and tools to support the process and to provide training and coaching to employees. Some of the characteristics of a usable formal method are:

- A formal method should be intuitive.
- It should have tool support.
- A formal method should be teachable.
- It should be able to adapt to change.
- The technology transfer path should be defined.
- A formal method should be cost-effective.

For more information on formal methods, see O'Regan (2017b).

Chapter 28
GPS

The Global Positioning System (GPS) is a network of satellites that orbit the Earth (Fig. 28.1). It is owned by the US government and operated by the US Air Force, and this global satellite system is used for civilian, commercial, and military purposes. It provides real-world geographic information including location (three-dimensional position), timing, velocity, and navigation information to users who are equipped with a GPS receiver (e.g., a mobile phone).

It may be used anywhere in the world irrespective of weather conditions, where there is an unobstructed line of sight of at least four GPS satellites. This free and dependable service operates independently of Internet or telephone reception, and the user is not required to transmit any data.

GPS consists of three segments, namely, the *space segment*, the *control segment*, and the *user segment*. The GPS space segment consists of a constellation of satellites orbiting in six planes and transmitting radio signals to users. The GPS satellites fly in a medium Earth orbit (MEO) at an altitude slightly over 20,000 km, and each satellite circles the Earth twice a day.

The GPS control segment consists of a global network of ground facilities that track the GPS satellites, monitors their transmissions, and performs appropriate analysis and management of the constellation. It includes a master control station, an alternate master control station, command and control antennae, and monitoring sites.

The GPS user segment consists of the GPS receiver equipment, which uses the signals and transmitted information from the GPS satellites to determine the user's three-dimensional position and time. These devices contain small computers that measure the time that it takes for the radio signal to travel from a GPS satellite until it arrives at the GPS antenna. The GPS receiver software calculates its position through a process of triangulation (Fig. 28.2).

The importance of being able to locate one's position on the surface of the Earth has been recognized for hundreds of years, and early navigation systems used celestial observation, spherical trigonometry, and hand computation. Electronic navigation commenced in the 1940s with fixed land-based radio transmitters.

© Springer Nature Switzerland AG 2018
G. O'Regan, *The Innovation in Computing Companion*,
https://doi.org/10.1007/978-3-030-02619-6_28

Fig 28.1 GPS Satellite
System (24 satellites).
Creative Commons

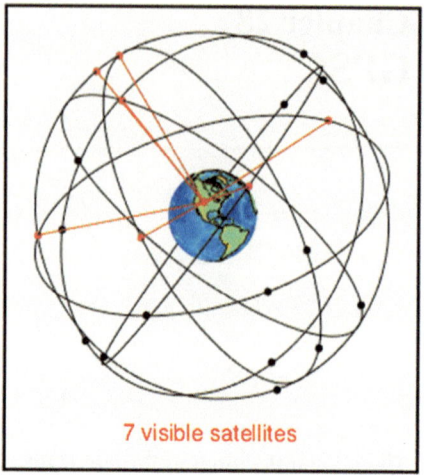

Fig 28.2 GPS in operation

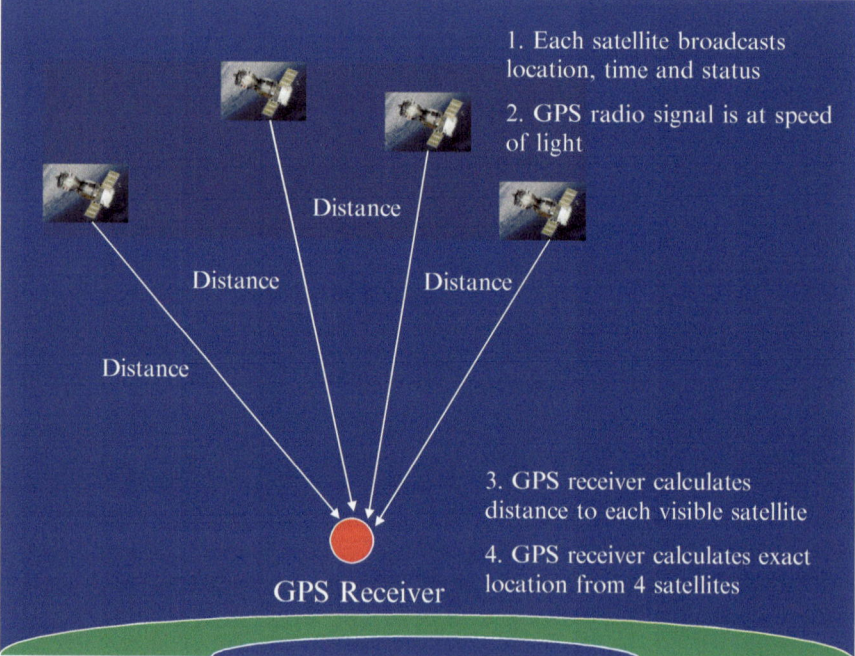

The first satellite navigation system (the TRANSIT system) was successfully tested in the United States in 1960, and it used a constellation of five satellites that could provide a navigation fix roughly once per hour. The GPS project commenced in the late 1970s, and it was designed to overcome the limitations of earlier navigation systems.

The GPS was developed for the US Department of Defense, and its inventors are Roger Easton of the Naval Research Laboratory, Ivan Getting of the Aerospace Corporation, and Bradford Parkinson of the Applied Physical Laboratory. It initially used a constellation of 24 satellites with 4 satellites in each of 6 orbits. The GPS became operational in 1995, and the orbits are evenly spaced every 60° around the Earth and at an altitude of 20,200 km. Several of the satellites are visible from any point on the surface of the Earth at any given time.

There are also several other satellite navigation systems under development or in use. These include the Russian Global Navigation Satellite System (GLONASS), the Chinese BeiDou Navigation Satellite System (BDS), and the European Galileo Global Navigation Satellite System (GNSS). Galileo was designed purely for civilian use, and it currently has early operational capability. The complete 30-satellite Galileo system (24 operational and 6 spares) is expected to be fully operational by 2020.

The typical handheld GPS receivers are accurate to about 5 m, and there are more sophisticated GPS receivers that provide position accuracy to within a centimeter.

28.1 Basic Principles of GPS

Figure 28.2 illustrates the basic operation of the GPS, where the distance is calculated based on the time that it takes a radio wave to travel from the satellite to the receiver or from the satellite to the control segment (i.e., $d = ct$, where d is the distance traveled, c is the speed of light, and t is the transmission time). The accuracy of the GPS reading (in terms of the latitude and longitude position) is improved when the GPS receiver can see additional satellites.

The GPS receiver must have an unobstructed view of the satellite, and so GPS does not work inside buildings, inside caves, or in dense forests. There may be interference to the signal from Earth's atmosphere, solid structures, and electromagnetic fields. Normally, six satellites are in view of any location at a given time, provided that the view is not obstructed. Satellites continuously transmit their location and time data.

The GPS receiver allows the user's location to be pinpointed in terms of latitude and longitude coordinates. Further, it provides additional information including the user's speed and altitude, as well as the direction in which the user is heading.

We illustrate the basic idea of determining a person's location in two dimensions given its distance from three known points A, B, and C in the following example (this is a simplification of what GPS does to determine a person's location in three dimensions). We may determine the person's location in two dimensions from the following information (Fig. 28.3):

(1) x is 500 km from point A.
(2) x is 475 km from point B.
(3) x is 450 km from point C.

Fig 28.3 Position at distance from a known point

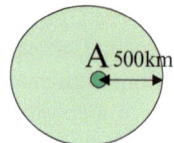

Fig 28.4 Position at distance from two known points

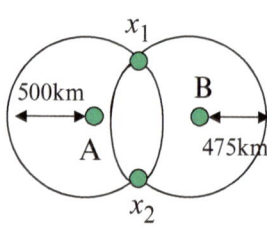

Fig 28.5 Position of x at distance from three known points

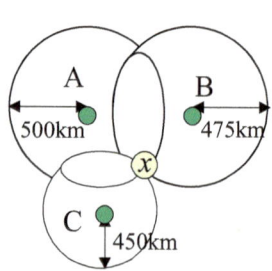

The location of x is determined in several steps, and from step 1, we note that x is 500 km from A, and so we draw a circle with radius 500 km with A at its center, and so we know that x is some point on the circumference of the circle (Fig. 28.4).

Step 2 gives additional information that x is 475 km from B, and so we draw a second circle with radius 475 km and with B at its center. We note that this circle intersects the first circle at point x_1 and x_2, and so we now know that x's position is either x_1 or x_2.

Finally, we use the information that x is 450km from C to draw a third circle with radius 450 km and with C at its center. We note the point of intersection of all three circles, and this is x's position (Fig. 28.5).

We have presented the solution visually to illustrate what is happening, and in practice, algebra would be used to find the solution.

28.2   Applications of GPS

GPS was originally developed for US military use during the cold war, with development commencing in the late 1970s. It was released for civilian and commercial use from the late 1990s. The free and dependable nature of GPS has led to the development of many applications of the technology from mobile phones, to

Table 28.1 Applications of GPS

Area	Description
Military	GPS was originally developed for the US military, but it has since been adopted by the armed forces of several countries around the world. It is used to map the location of vehicles and other assets on battlefields in real time. GPS is also fitted to missiles for tracking and guidance
Road transport	GPS is used for commercial fleet management and taxi services. The emergency services and private motorists use in-car navigation systems, and GPS plays an important role in the navigation of driverless cars
Aviation	Almost all modern aircraft are equipped with several GPS receivers that provide real-time aircraft position and a map of flight progress. Unmanned aerial vehicles also use it for navigation
Maritime	High-accuracy GPS receivers allow the captain to navigate safely through unfamiliar channels or waterways
Surveying and mapping	Surveyors are responsible for accurate mapping and measuring features of Earth's surface (e.g., land boundaries, mapping sea floor). Accurate GPS receivers make this easier, and the data can be transferred into mapping software to create quick and detailed maps
Telecommunications	GPS provides the facility for the synchronization of coordinated universal time (UTC)
Social	GPS is used for various social activities such as hiking, sailing, and geotagging photographs

wristwatches, to the transport sector including vehicles and aviation, to outdoor pursuits, and to surveying and mapping. Today, millions of users employ satellite navigation to travel from A to B, as well as using many other applications of the technology.

A GPS receiver allows the user to accurately pinpoint their position and velocity in vehicles, aircraft, and ships. Drivers can use in-vehicle navigation systems to follow a route and to avoid traffic problems. GPS can be used by hikers to check that they are following their chosen route, and GPS is invaluable to the emergency services in enabling them to find their way to an incident faster, as well as pinpointing the location of accidents for the search and rescue teams. GPS also supports the creation of accurate maps. Table 28.1 lists a small number of applications of GPS.

There are many texts available (e.g., see Misra and Enge (2010)) that provide more detailed information on GPS.

Chapter 29
Graphical User Interface and Human-Computer Interaction

Human-computer interaction (HCI) is a branch of computer science that is concerned with the design, evaluation, and implementation of interactive computing systems for human use. It is focused on the interfaces between people and computers and involves several fields including computer science, cognitive psychology, design, and communication. The human-computer interaction field has grown over the decades from batch systems to text-based interaction systems, to graphical user interfaces (GUI), and to voice user interfaces (VUI) for speech recognition and speech synthesis.

The success of computer systems is critically influenced by the design of the human-computer interaction and in the achievement of end-user computing satisfaction. Human-computer interaction is concerned with the study of humans and machines, and so it needs knowledge of both to be effective. The study of machines requires knowledge of computer graphics, programming languages, capabilities of current technology, and so on, whereas on the human side, it requires knowledge of cognitive psychology, ergonomics, and other human factors such as usability and computer user satisfaction.

There are several fundamental principles and models underlying HCI. It is essential to understand the users and their characteristics, as well as their diversity in age, experience, physical and intellectual abilities, and so on. It is customary to distinguish between two types of user knowledge (IT and domain knowledge), and the user's proficiency in each type of knowledge yields several user categories that range between novice and expert.

- Interface knowledge (knowledge of the IT technology).
- Domain/task knowledge of the real-world system.

Usability is important in software engineering, and especially since the emergence of the World Wide Web in the early 1990s. The usability of the software is the perception that a user or group of users has of its quality and ease of use (i.e., is the software easy to use and easy to learn?), and its efficiency and effectiveness. Usability is a multidisciplinary field, and psychological testing may be employed to

© Springer Nature Switzerland AG 2018
G. O'Regan, *The Innovation in Computing Companion*,
https://doi.org/10.1007/978-3-030-02619-6_29

evaluate the perception that users have of the computer system. Usability is the extent to which software may be used to achieve the customers' objectives efficiently.

User-centered design (UCD) is a design process that is focused on the usability of and accessibility of the system to be developed, and it places the user at the center of the software development process. The users are actively involved from the beginning of the project, and they provide regular feedback at each stage of the process. UCD follows well-established techniques for analysis and design, and it is focused on understanding the characteristics of the users and their needs.

Information technology professionals and computer hobbyists were the only humans that interacted with computers in the early days of the computing field up to the mid-1970s. This changed following the invention of the microprocessor in the early 1970s, which led to the development of home and person computers from the mid-1970s/early 1980s. It meant that everyone in the world was now a potential computer user, and it was clear that there were serious deficiencies with respect to the usability of computers as tools.

Humans interact with computers in many ways, and so it is important to understand the interface between them to enhance their interaction. The early interaction between humans and machines was via *batch processing* (running programs in batches without human intervention) on large expensive mainframe computers. The interaction between the human (operator) and computer was limited and consisted of placing the punched cards (encoded instructions to the computer) on the card reader, and the computer would then process the cards overnight. These early computers were slow and expensive, and it was essential that they were used efficiently 24 h a day. Most of these early computers could run only one program at a time, and programmers were unable to interact with the computer while it was running. This meant that it was difficult and time-consuming to identify and correct errors.

Licklider wrote an influential paper *Man-Computer Symbiosis* in 1960, in which he outlined the need for a simple interaction between users and computers (Licklider 1960). This paper mentioned novel ideas such as sharing computers among many users, interactive information processing and programming, large-scale storage and retrieval, and speech and handwriting recognition.

Doug Engelbart and others at the Augmentation Research Centre developed the revolutionary NLS (On Line System) computer system in the late 1960s (see Chap. 40). This online word processor system introduced novel features such as the first computer mouse, hypertext links, time-sharing, and a command line interface. User trials and testing were employed in its development as part of a *philosophy toward a system adapting to people rather than people adapting to a system.*

A *text-based interface* (also known as a command line interface) is where the system interaction (input and output) and navigation is text-based. They are easier to use than punched card programming, but they are not intuitive and required effort from the user in remembering long lists of system commands.

One of the most famous text-based operating systems was Microsoft's MS/DOS operating system for IBM compatible personal computers (Fig. 29.1), which was introduced in 1981 (see Chap. 42). Text-based interfaces are effective for expert

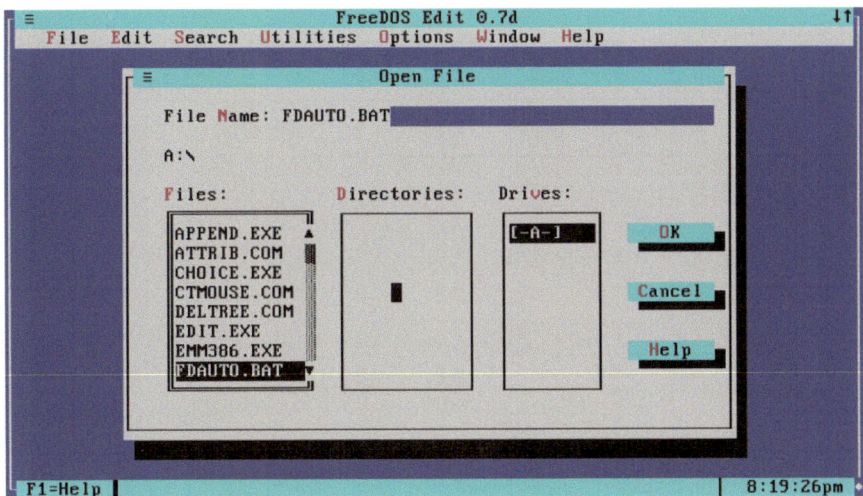

Fig. 29.1 FreeDOS text editing

users, but they are more difficult for users with an average level of knowledge. They have a steep learning curve, as it is difficult to remember a long list of system commands, and the fact that they are not very intuitive or user-friendly motivated research into other forms of human-computer interaction.

The *graphical user interface* (GUI) is a human-computer interface that uses graphical icons, menus, and windows to represent information and action to the user. They are intuitive and user-friendly and have revolutionized human and computer interaction. They have made computers and electronic devices attractive to nontechnical users, and the usability of the GUI has allowed many users with varying ability and expertise to successfully interact with computers.

Early work on graphical user interfaces took place at Xerox PARC in the 1970s with their work on the Xerox Alto personal workstation. This was the first computer to use a mouse-driven graphical user interface, and it was introduced in the mid-1970s. The Alto workstation was essentially a small minicomputer rather than a personal computer (it was not based on the microprocessor). Its significance is that it had a major impact on future user interface design and especially on the design of the Apple Macintosh computer.

The Xerox Star was introduced in the early 1980s, and it followed sound usability principles (prototyping and analysis, iterative development, and testing with users) in its development. Steve Jobs visited Xerox PARC in the late 1970s to see their pioneering work on the Xerox Alto, and he immediately recognized that the future of personal computing was with computers that employed a GUI. The design of the Apple Macintosh was heavily influenced by the design of the Xerox Alto, and the release of the Apple Macintosh was a major milestone in computing (see Chap. 7).

The Macintosh was a much easier machine to use than the existing IBM personal computer. Its friendly and intuitive graphical user interface was a revolutionary

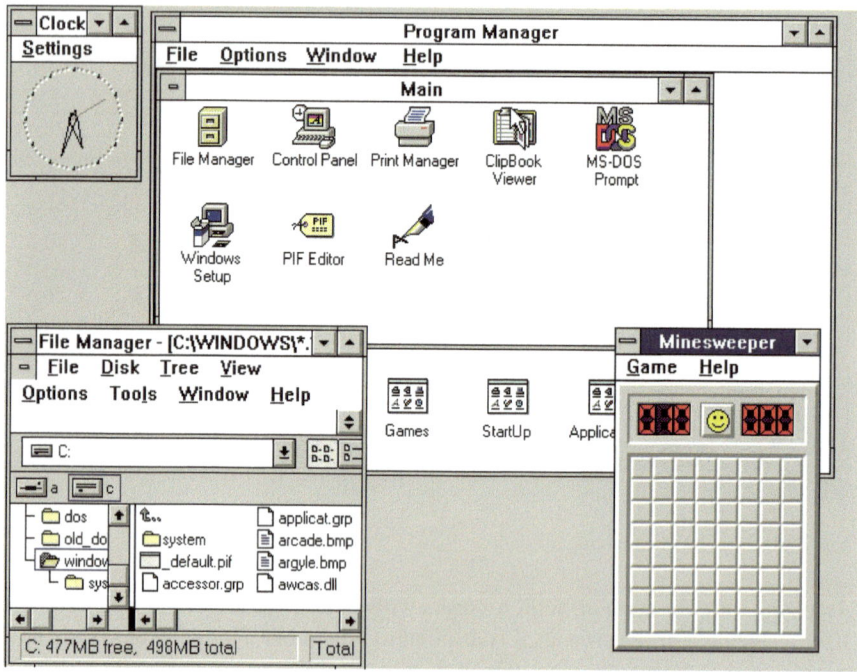

Fig. 29.2 Microsoft Windows 3.11 (1993). Used with permission from Microsoft

change from the command-driven operating system of the IBM PC, which required the users to be familiar with its operating system commands. It was in 1990 before Microsoft introduced its Windows 3.0 GUI-driven operating system (Fig. 29.2).

Today, the prevalent paradigm in human-computer interaction is the WIMP (windows, icons, menus, and pointers) paradigm, which is comprised of a graphic and text interface navigated by a mouse and keyboard (i.e., a GUI). The future of HCI is predicted to be the SILK (Speech, Image, Language, Knowledge) paradigm, where communication between humans and machine will be more natural and intuitive. It will involve voice recognition systems with voice output capabilities, as well as natural language processing and artificial intelligence capabilities.

Chapter 30
Harvard Mark 1 Computer

The Harvard Mark I was a large machine designed to assist in the numerical computation of differential equations. It was designed by Howard Aiken at Harvard University, and the machine was funded and built by IBM. The machine was known as the Harvard Mark I (it was also known as the *IBM Automatic Sequence Controlled Calculator* (ASCC)). Aiken made several important contributions to the early computing field, and he demonstrated that a large calculating machine could be built that would provide speedy solutions to mathematical problems. He also designed successor machines to the Mark I, and these include the Mark II, III, and IV, as well as making important contributions to early computer science education.

Aiken (Fig. 30.1) was born in New Jersey in 1900, and he obtained a bachelor's degree in electrical engineering from the University of Wisconsin in 1923. He worked in industry for several years and returned to academia in the 1930s. He was awarded a PhD degree in Physics from Harvard in 1939.

He became conscious during his graduate studies of the need for a machine that could deal with many of the tedious calculations that arose in solving differential equations by numerical means. This motivated his research into machines that could ease the burden of calculation.

Aiken considered what a scientific calculating machine should do, and he published a report to communicate his ideas. His goal was to construct an electromechanical machine that could perform mathematical operations quickly and efficiently, and the machine would need to be able to handle positive and negative numbers, scientific functions such as logarithms, and be able to work with minimal human intervention.

He discussed the idea with colleagues and with IBM, and he was successful in obtaining IBM funding to build the machine. The machine was built at the IBM laboratories at Endicott with several IBM engineers involved in its construction. Aiken was influenced by Babbage's ideas on the design of the Analytic Engine, and the machine was able to compute mathematical tables using the method of finite differences as envisaged by Babbage (see Chap. 6). The construction took 7 years and was completed in 1943, and IBM presented the machine to Harvard University

© Springer Nature Switzerland AG 2018
G. O'Regan, *The Innovation in Computing Companion*,
https://doi.org/10.1007/978-3-030-02619-6_30

Fig. 30.1 Howard Aiken

in 1944. The Harvard Mark I was essentially an electromechanical calculator that could perform large computations automatically. It could perform operations such as addition, subtraction, multiplication, and division as well as referring to previous results.

It was designed to assist in the numerical computation of differential equations, and it was 50 feet long and 8 feet high and weighed 5 tons. It performed additions in less than a second, multiplications in 6 s, and division in about 12 s. It used electromechanical relays to perform the calculations (Fig. 30.2).

It could execute long computations automatically. It used 500 miles of wiring and had over 700,000 components. It was the industry's largest electromechanical calculator, and it had 60 sets of 24 switches for manual data entry. It could store 72 numbers, each 23 decimal digits long. Punched cards were used to input the data, and the results were either on punched cards or an electric typewriter.

The US Navy used it for ballistic calculations, and it remained in use until 1959. The machine played a role during the Manhattan project (the project to develop the Atomic bomb) in running various problems related to implosion. The machine cost approximately half a million dollars, but it was never mass produced by IBM.

The announcement of the Harvard Mark 1 led to tension between Aiken and IBM, as Aiken announced himself as the sole inventor without acknowledging the important role played by IBM in funding and building the machine. Aiken also designed and developed the Harvard Mark II, III, and IV. The Mark II was funded by the US Navy and completed in 1947. It was faster than the Mark I and employed an electrical memory.

His Mark III (or Aiken Dahlgren Electronic Calculator (ADEC)) was built in 1949 for the Navy, and it was one of the earliest machines to use magnetic drums for

The Harvard Mark I

Fig. 30.2 Harvard Mark 1 (Courtesy of International Business Machines Corporation, International Business Machines Corporation)

data storage. His Mark IV was completed in 1952 with funding from the US air force.

Several terms used in computer programming originated with the Mark I. These include terms such as *patch*, *bug*, and *library*. The term "patch" referred to small corrections in the program sequence by patching over portions of the paper tape and re-punching the holes in that section. The term "bug" was coined by Grace Murray Hopper to refer to unexpected problems that arise during the coding. Hopper identified the origin of a famous bug to a moth stuck in one of the relays of the Mark II computer in 1947. The term "library" referred to sections of tape from previous problems that are cut, stored, and pasted back together for new uses and forms a repository of computer code.

Aiken also made important contributions to computer science education, and he started one of the earliest computer science academic programs in the world at Harvard. He received the IEEE Edison Medal in 1970 for his pioneering contributions to the development and applications of large-scale digital computers and to important contributions to education in the computer field.

Chapter 31
Hollerith's Tabulating Machines and the Birth of IBM

Hollerith's punch card tabulating machine was designed to process the results of the 1890 population census in the United States. It used an electric current to sense holes in punched cards, and it kept a running total of the data. This allowed the statistics to be recorded by electrically reading and sorting punched cards. The new methodology enabled the results of the 1890 census to be available in a couple of months rather than years, and it led to a saving of millions of dollars in tabulating the census data.

Hermann Hollerith (Fig. 31.1) was an American statistician, inventor, and entrepreneur. He qualified as a mining engineer from Columbia University School of Mines in 1879, and he became an assistant to one of his teachers, Dr. William Trowbridge, after his graduation.

Trowbridge was appointed as a consultant to the US Census Board, and Hollerith joined the Census Board as a statistician. His goal was to assist the Census Board in processing the large amount of data generated from the 1880 population census[1] of the United States. This stimulated his interest in automating the tabulation of the census data and in manipulating the data mechanically.

Dr. John Billings[2] believed that there must be a mechanical way to do the tabulation, and he communicated his ideas with Hollerith. Billings believed that the holes in the card should be able to tabulate the population and similar statistics in a similar way to that done by a card in the Jacquard loom that regulates the pattern to be woven in the fabric.

Hollerith examined the workings of the Jacquard loom to see if it could assist in the tabulation. The main feature that he considered applicable was its use of punched cards, as these are an efficient way of storing information. He also observed that it should be possible to punch information onto a card, as a conductor might do on a

---

[1] The processing of the 1880 population census was done manually and took several years to complete.

[2] Dr. John Billings did pioneering work on statistics for the 1880 and 1890 population census, and he was also a distinguished surgeon.

© Springer Nature Switzerland AG 2018   151
G. O'Regan, *The Innovation in Computing Companion*,
https://doi.org/10.1007/978-3-030-02619-6_31

463-465 PENNA. AVENUE,
WASHINGTON, D. C.

train. Hollerith conducted several experiments and initially employed a paper tape rather than cards. A pin could go through a hole in the tape and complete an electrical circuit, and Hollerith later replaced paper with cards as these offered a more effective solution.

He took a position at the US Patent Office in 1884, and he patented his invention later that year (Fig. 31.2). He understood the importance and potential value of protecting his inventions, and he was granted a total of 30 patents during his career. He did further work on methods to convert the information on punched cards into electrical impulses, which could activate mechanical counters. He initially employed the ticket punch used by conductors on the railway to punch holes, but he later designed a more effective punch for his system.

Hollerith's system was first tested on mortality systems in Baltimore in 1887. The expectation was that the 1890 population census would take several years to process, and the US Census Bureau recognized that its existing methodology was no longer fit for purpose. It held a contest to find a more efficient and cost-effective solution, and Hollerith's system was the clear winner.

His punch card tabulating machine used an electric current to sense holes in punched cards, and it kept a running total of the data. This allowed the statistics to be recorded by electrically reading and sorting punched cards. The new methodology enabled the results of the 1890 census to be available in a couple of months rather than years, and this led to a saving of millions of dollars in tabulating the census data. The US population was recorded to be over 62 million in 1890.

Fig. 31.2 Hollerith's Tabulator (1890). (Courtesy of Courtesy of International Business Machines Corporation, © International Business Machines Corporation)

Hollerith's system was later used to tabulate the census data in several other countries including Russia and Canada.

Hollerith formed the Tabulating Machine Company in Washington, D.C. in 1896, and this was the first electric tabulating machine company. Hollerith's company merged with the International Time Recording Company to form the Computing Tabulating Recording Company (CTR) in 1911. Thomas Watson joined the company in 1914 when the company was going through difficulties. He turned the company around, and the company changed its name to *International Business Machines* (*IBM*) in 1924. IBM has been in business for over 100 years, and it is a respected leader in the computing field.

Hollerith's punched cards and tabulating machine were a step forward toward automated computation. His device could automatically read information that had been punched on to the card, but the machine was limited to tabulation and could not be used for more complex computations. The tabulator was an important precursor to the modern computer, and punched card technology remained in use on computers up to the late 1970s.

Chapter 32
Integrated Circuit

The electronics industry was dominated by vacuum tube technology up to the mid-1950s. However, vacuum tubes had inherent limitations as they were bulky, unreliable, produced considerable heat, and consumed a lot of power. Bell Labs invented the transistor in the late 1940s (see Chap. 52), and transistors were tiny and consumed very little power. Further, they were more reliable and lasted longer than the bulky vacuum tubes.

The transistor stimulated engineers to design ever more complex electronic circuits and equipment containing hundreds or thousands of discrete components such as transistors, diodes, rectifiers, and capacitors. Each component needed to be wired to many other components, and the wiring and soldering was done manually. Clearly, more components would be required to improve performance, and therefore it seemed that future designs would consist almost entirely of wiring.

The problem with this was that these components needed to be interconnected to form electronic circuits, and this involved hand soldering of thousands of components to thousands of bits of wire. This was expensive and time-consuming, and every soldered joint was a potential source of trouble leading to problems with the reliability of the circuit. The challenge for the industry was to find a cost-effective and reliable way of producing these components and interconnecting them. The invention of the integrated circuit was the solution to the problems that engineers faced in increasing the performance of their designs as the number of components in the design increased.

Jack Kilby (Fig. 32.1) joined Texas Instruments in 1958, and he began investigating how to solve this problem. He realized that semiconductors were all that were really required, as resistors and capacitors could be made from the same material as the transistors. He realized that since all of the components could be made of a single material, they could also be made in situ interconnected to form a complete circuit.

Kilby succeeded in building an integrated circuit made of germanium that contained several transistors in 1958. Robert Noyce of Fairchild Semiconductors built an integrated circuit on a single wafer of silicon in 1960, and Kilby and Noyce are

© Springer Nature Switzerland AG 2018
G. O'Regan, *The Innovation in Computing Companion*,
https://doi.org/10.1007/978-3-030-02619-6_32

Fig 32.1 Jack Kilby.
(Courtesy of Texas
Instruments)

Fig 32.2 First integrated
circuit. (Courtesy of Texas
Instruments)

considered co-inventors of the integrated circuit. Kilby was awarded the Nobel
Prize in Physics in 2000 for his role in its invention.

Kilby's integrated circuit consisted of a transistor and other components on a
slice of germanium (Fig. 32.2). His invention revolutionized the electronics industry, and the integrated circuit is the foundation of almost every electronic device in
use today. The size of the first integrated circuit was 7/16 by 1/16 inches, and it was
made out of germanium rather than silicon.

Robert Noyce at Fairchild Semiconductors developed an integrated circuit based
on a single wafer of silicon in 1960, and today silicon is the material of choice for
semiconductors. Noyce made an important improvement on Kilby's design in that
he added a thin layer of metal to the chip to better connect the various components

in the circuit. Noyce's solution made the integrated circuit more suitable for mass production, and Fairchild Semiconductors pioneered the use of the *planar process* for making transistors, and the existing semiconductor companies soon employed this process.

An *integrated circuit* (IC) consists of a set of electronic circuits on a small chip of semiconductor material, and it is much smaller than a circuit made from independent components. The IC is made on a small plate of semiconductor material that is usually made of silicon. It is extremely compact and may contain billions of transistors and other electronic components in a tiny area. The width of each conducting line has got smaller and smaller due to advances in technology over the years, and it is now measured in tens of nanometers.[1] The invention of the integrated circuit led to major reductions in the size and cost of making electronics, and it impacted the design of all future computers and other electronics.

The size of the components in a modern fabrication plant is extremely small, with thousands of transistors fitting inside the cross section of a strand of hair. The production of a chip requires precision at the atomic level, with tiny particles such as those in tobacco smoke large enough to ruin a chip. For this reason, chip production takes place in a *clean room*, which is a special room designed with furniture made of special materials that don't give off particles, and very effective air filters and air circulation systems.

There has been a massive reduction in the production costs of integrated circuits, with the initial production cost of an IC at $1000 in 1960. However, as demand increased and production techniques improved, the cost of production was reduced to $25 by 1963.

There are several generations of integrated circuits from the small-scale integration (SSI) of the early 1960s, which typically had less than 30 transistors on the chip, to medium-scale integration (MSI) of the late 1960s with less than 300 transistors on the chip; to large-scale integration (LSI) of the mid-1970s with less than 3000 transistors on the chip; and to very large-scale (VLSI) and ultra large-scale integration (ULSI) of the 1980s, which have over a million transistors on the chip.

There are several large companies that design and make semiconductors, including Texas Instruments (TI), which is an American electronics company that is one of the largest manufacturers of semiconductors in the world. Intel and AMD (Advanced Micro Devices) are among the largest makers of semiconductors in the world.

32.1   Moore's Law

Gordon Moore observed that over a period of time (from 1958 up to 1965), the number of transistors on an integrated circuit doubled approximately every year. This led him to formulate what became known as *Moore's Law* in 1965 (Moore

---

[1] 1 nanometer (nm) is equal to 10^{-9} m.

1965), which predicted that this trend would continue for at least another 10 years. He refined the law in 1975 and predicted that a doubling in transistor density would occur every 2 years for the following 10 years.

His prediction of *exponential growth* in transistor density has proved to be accurate over the last 50 years, and the significant growth in capability of modern digital electronic devices is linked to Moore's Law.

The exponential growth in processor speed and memory size is directly related to this law. However, it is likely that the growth in transistor density will slow down significantly in the future due to the limitations in miniaturizing transistors. The phenomenal growth in productivity is due to continuous innovation and improvements in manufacturing processes. It has led to more and more powerful computers running more and more sophisticated applications.

However, it remains an open question as to whether the exponential increase in computational power can be extended in the future due to the limitations of Moore's Law as the semiconductor industry is approaching the atomic level in miniaturization. It is possible that other technologies such as nanotechnology or quantum computing may offer other solutions that will deliver increases in computational power to deal with limitations of Moore's law when it is one atom per memory cell.

32.2   Early Integrated Circuit Computers

It took some time for integrated circuits to take off, as they were an unproven technology, and they remained expensive until mass production. Texas Instruments commercialized the integrated circuit by designing a handheld calculator that was as powerful as the existing large, electromechanical desktop models. The resulting electronic handheld calculator was small enough to fit into a coat pocket. It was a battery-powered device and could perform the four basic arithmetic operations on six-digit numbers. It was completed in 1967.

The earliest computers with integrated circuits appeared in the 1960s, and their early use was mainly in embedded systems. Integrated circuits played an important role in early aerospace projects such as the Apollo Guidance Computer and Minuteman missile. The Apollo flight computer was one of the earliest computers to use integrated circuits, and it was developed by MIT/Raytheon and introduced in 1966. It provided capabilities for the guidance, navigation, and control of the Apollo spacecraft. The Minuteman II program used a computer built from integrated circuits, and the guidance system of the Minuteman II intercontinental ballistic missile was much smaller due to their use.

Fig 32.3 PDP – 8/e

DEC's first minicomputer to use integrated circuits was the popular PDP-8 (Fig. 32.3), which was designed by Edson de Castro and introduced in 1965. Hewlett-Packard introduced the 2116A minicomputer in 1966, and this machine used Fairchild Semiconductors integrated circuits.

The Honeywell ALERT airborne computer was designed to handle complex airborne data in a real-time environment, and it was introduced in 1966. The Central Air Data Computer was designed in the late 1960s, and it was used for flight control in the US Navy's F-14A Tomcat Fighter.

Chapter 33
Internet

The computers in the 1960s were large expensive machines with limited processing power, and communication between them was virtually nonexistent. Several computer scientists had dreams of a worldwide network of computers, with every computer around the globe connected to all others. This would allow everyone in the world to be connected and to access programs and data from anywhere.

The US Department of Defense founded the Advanced Research Projects Agency (ARPA) in the late 1950s, as a body to manage the development of new and advanced technologies for the US military. The concept of packet switching[1] was invented in the early 1960s, and several organizations began work on its implementation.

The first (nonmilitary) wide-area network connection was created in 1965 and involved the connection of a computer at MIT to a computer in Santa Monica (via a dedicated telephone line). This demonstrated the feasibility of a telephone line for data transfer, and ARPA recognized the need to build a network of computers. This led to the ARPANET project in 1966, which aimed to implement a packet-switched network with a network speed of 56 Kbps. The ARPANET packet switching network was introduced in the late 1960s, and it remained operational until 1990.

ARPA was renamed to DARPA in 1972, and it commenced a project to connect several computers in diverse geographical locations using a radio-based network and a project to establish a satellite connection between a site in Norway and in the United Kingdom. This led to the requirement of the interconnection of ARPANET with other networks. The key problems were to investigate ways of achieving convergence between the ARPANET, radio-based networks, and the satellite networks, as these all had different interfaces, packet sizes, and transmission rates. Therefore, there was a need for a network-to-network connection protocol.

The concept of the Transmission Control Protocol/Internet Protocol (TCP/IP) was developed at DARPA by Vint Cerf (Fig. 33.1) and Bob Kahn in 1974 (Kahn and Cerf 1974). TCP is a set of network standards that specify the details of how

---

[1] Packet switching is a message communication system between computers. Long messages are split into packets that are then sent separately to minimize the risk of congestion.

© Springer Nature Switzerland AG 2018
G. O'Regan, *The Innovation in Computing Companion*,
https://doi.org/10.1007/978-3-030-02619-6_33

Fig. 33.1 Vint Cerf

Table 33.1 TCP/IP layers

Layer	Description
Network interface layer	This layer is responsible for formatting packets and placing them onto the underlying network
Internet layer	This layer is responsible for network addressing. It includes the Internet Protocol and the Address Resolution Protocol
Transport layer	This layer is concerned with data transport and is implemented by TCP and the user datagram protocol (UDP)
Application layer	This layer is responsible for liaising between user applications and the transport layer. It includes the File Transfer Protocol (FTP), telnet, Domain Name System (DNS), and Simple Mail Transfer Protocol (SMTP)

computers communicate, as well as the standards for interconnecting networks. It allows the inter-networking of very different networks, which then act as a single network.

TCP/IP was designed to be flexible, and it provides a transmission standard that deals with physical differences in host computers, routers, and networks. It is designed to transfer data over networks which support different packet sizes and which may sometimes lose packets.

The TCP/IP protocol standards are the *Transmission Control Protocol* (TCP) and the *Internet Protocol* (IP). TCP details how information is broken into packets and reassembled on delivery, whereas IP is focused on sending the packet across the network. The TCP/IP standards allow users to send electronic mail or to transfer files electronically, without needing to concern themselves with the physical differences in the networks. TCP/IP consists of four layers (Table 33.1).

The Internet Protocol (IP) is a connectionless protocol that is responsible for addressing and routing packets. It breaks large packets down into smaller packets when they are traveling through a network that supports smaller packets. A *connectionless* protocol means that a session is not established before data is exchanged, and packet delivery with IP is not guaranteed as packets may be lost or delivered out of sequence. An acknowledgment is not sent when data is received, and the sender or receiver is not informed when a packet is lost or delivered out of sequence.

The router forwards a packet only if it knows a route to the destination, and otherwise the packet is dropped. Packets are dropped if their checksum is invalid or if their time to live is zero. The acknowledgment of packets is the responsibility of the TCP protocol. TCP/IP is the fundamental protocol at the heart of the Internet, and ARPANET employed the TCP/IP protocols as a standard from 1983.

Cerf and Kahn have received numerous awards for their contributions to the development of the Internet. These include the US National Medal of Technology and Innovation, which they received from President Clinton in 1997. They received the ACM Turing Award in 2004 for their work on Internet Protocols.

33.1   Birth of the Internet

The use of ARPANET was initially limited to academia and to the US military, and there was little interest from industrial companies in the early years. The network allowed messages to be sent between the universities that were part of ARPANET, and there were over 2000 hosts on the TCP/IP-enabled network by the mid-1980s.

It was decided to shut down the network by the late 1980s, and the National Science Foundation (NSF) commenced work on its successor, the NSFNET, in the mid-1980s. This network consisted of multiple regional networks connected to a major backbone. The original links in NSFNET were 56K bps, but these were updated to 1.544 Mbps T1 links in 1988. The NSFNET T1 backbone initially connected 13 sites, but this increased, as there was growing academic and industrial interest from around the world. The NSF quickly realized that the Internet had commercial potential.

The Internet began to become more international with nodes in Canada and several European countries. DARPA formed the Computer Emergency Response Team (CERT) to deal with any emergency incidents arising from the operation of the network.

The independent not-for-profit company, Advanced Network Services (ANS), was founded in 1991. It installed a new network (ANSNET) that replaced the NSFNET T1 network, and it operated over T3 (45 Mbps) links (Fig. 33.2). It was owned and operated by a private company rather than the US government, with the NSF focusing on the research aspects of networks rather than on the operational side.

The ANSNET network was a distributive network architecture operated by commercial providers such as Sprint, MCI, and BBN. The various parts of the network were connected by major network exchange points. These were termed Network Access Points (NAPs). There were over 160,000 hosts connected to the Internet by the late 1980s, and the early commercial use of the Internet in the late 1980s included

NSFNET T3 Network 1992

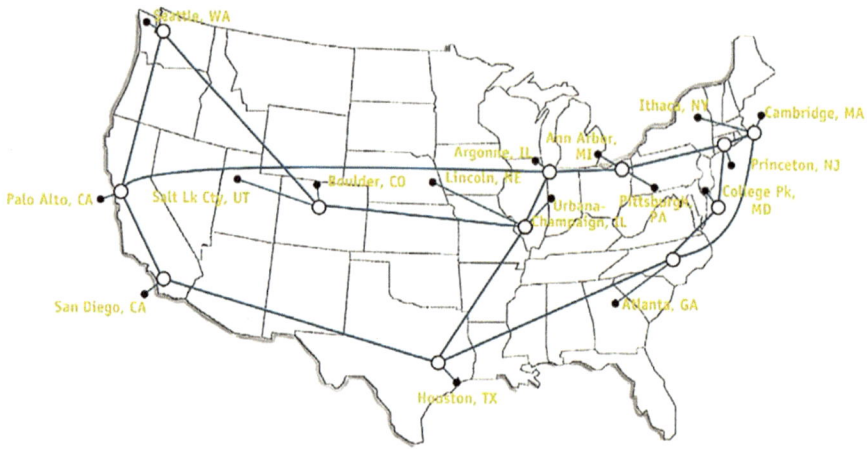

Fig. 33.2   NSFNET T3 network – 1992. Public domain

email communication (e.g., CompuServe which provided an online chat system and message board) between the users of the Internet.

The use of the Internet exploded following Tim Berners-Lee's invention of the World Wide Web at CERN in the late 1980s (Berners-Lee 1999). Berners-Lee's built on existing inventions such as the Internet, hypertext, and the mouse to form the World Wide Web (see Chap. 57).

The Internet is a global network, which consists of many interconnected networks, and it operates without a single governing body. Internet service providers establish the worldwide connectivity between individual networks

The Internet has led to almost instantaneous communication, and it has led to electronic mail, the World Wide Web, social networking, electronic commerce, and telephone calls over the Internet. The web is a global set of documents and images referenced via universal resource locators (URLs) or hyperlinks. Internet telephony is a communication service made possible by the Internet, and it uses the Voice over Internet Protocol (VoIP). There are several billion Internet users in the world today.

33.2   Internet of Things

The Internet of Things refers to interconnected technology that is now an integral part of modern society, where computation and data communication are embedded in our environment. The Internet of Things is not a single technology as such, and instead it is a collection of devices, sensors, and services that capture data and are used to monitor and control the world around them. It means that information processing is now an integral part of people's lives.

The Internet of Things has been applied to several areas including our bodies (quantified self), our homes (smart homes), and public spaces (smart city). Wearable biometric sensors may be used to determine the calories burned during a period of exercise, as well as monitor heart rate, breathing, skin temperature, and perspiration. In theory, this helps individuals to control key parameters associated with their health.

The rise of the *smart home* is intended to deliver convenience to home occupiers, and these consist of connected devices that provide useful functions. These digital devices may be used to control lighting, entertainment, security, as well as cooling and ventilation systems in the modern home. They gather data about the environment as well as passing data back to the service provider. However, there are dangers with giving all this private data about your life to a service provider, and it is essential that the privacy of individuals be protected.

The rise of the *smart city* is where the modern city collects data about its residents and uses the data to make more efficient use of energy, space, and other resources. The data may be gathered through CCTV and other devices, and in the future the smart city will have knowledge of the habits and energy use of its citizens, allowing it to control resources more effectively.

There are several implicit assumptions with respect to smart cities, and it seems to be assumed that it is possible to know all aspects of the world perfectly with data, that the data will always be accurate, and that the data will be easy to interpret. These assumptions are questionable.

That is, while the Internet of Things presents many new possibilities, it is important to proceed cautiously and to use it as an extra tool that may support decision-making, rather than assuming that it provides all the answers. For further information on the Internet of Things, see the thought-provoking Guardian article (Greenfield 2017).

33.3   Internet of Money and Bitcoin

The idea of the Internet of Money is to build a financial environment that is suitable for the Internet world, and it moves away from the traditional centralized model where third-party banks record and manage all financial transactions. The new paradigm is that of a decentralized model via the Internet where buyers and sellers interact directly through *digital currencies* and decentralized ledgers. This new model is termed the *Internet of Money*, and *Bitcoin* aims to satisfy this decentralized model.

Digital currency is a type of currency that is available only in digital form, and it exhibits similar properties to traditional physical currencies in that it may be used to buy goods and services. It includes virtual currencies and cryptocurrencies.

The concept of digital cash was proposed by David Chaum in the early 1980s, and he formed DigiCash (an electronics cash company) in the early 1990s to commercialize his research (Chaum 1982). The goal of electronic cash (eCash) is to allow the user to be anonymous, and it allows users to spend in a manner that is untraceable to a bank or any other third party.

Chaum introduced the idea of blind signatures in his 1982 paper, which blinds the content of a message before it is signed. This means that the signer cannot determine the content of the message, but the resulting blind signature can be verified against the original unblinded message.

One of the earliest digital currencies was e-Gold (it was backed by Gold), and this centralized service appeared in the mid-1990s. The US government later shut it down due to concerns over money laundering. Q coins emerged around 2005, and Bitcoin appeared in 2008. Bitcoin is the most widely used and accepted digital currency, and it is based on cryptographic algorithms (i.e., it is a cryptocurrency).

There are several types of digital currency including centralized systems (e.g., PayPal, eCash) which sell digital currency directly to the end user, mobile digital wallets for contactless payment transfer to facilitate easy payment (e.g., Google Wallet and Apple Pay make it easy to carry all your debit and credit cards on your smartphone), and decentralized system which employ cryptocurrencies and rely on cryptography (Bitcoin is the most well-known of these). Finally, there are virtual currencies that are issued and controlled by its developers and accepted by the members of a virtual community.

Bitcoin is a cryptocurrency and digital payment system, and it is the first decentralized digital currency. It was created by an unknown inventor(s) with the pseudonym Satoshi Nakamoto in 2008 (Nakamoto 2008), and it works without a central repository or single administrator. It is peer to peer with transactions taking place directly between users without the need for third-party intermediaries, and the transactions are verified by network nodes and recorded in a public distributed ledger termed a blockchain. Nakamoto released the open-source software for Bitcoin in early 2009, and the domain name bitcoin.org was registered in 2008.

The unit of account in the Bitcoin system is the bitcoin (BTC) with smaller amounts represented by millibitcoins (0.001 BTC), and the smallest amount is the satoshi (0.00000001 BTC).

Chapter 34
Iridium System

The original Iridium was a global satellite phone company that was backed by Motorola, and in many ways, it was an engineering triumph over common sense. The Iridium system used up to over $5 billion in funds to build an infrastructure of low Earth orbit (LEO) satellites to provide worldwide mobile communication coverage. The satellites are at an altitude of approximately 780 km, and they have an approximate velocity of 27,000 km/h and complete an orbit of the Earth in under 2 h.

The 66 Iridium satellites are arranged in 6 orbits of 11 satellites each, and it typically takes each satellite about 10 min to cross the sky from horizon to horizon. The satellites are programmable, which allows engineers to upload software to keep the system functioning at the required performance and reliability levels. The system includes in-orbit spares that are in a storage orbit, and a spare can be quickly repositioned and activated in the event of failure.

The Iridium satellite constellation was conceived in the early 1990s to provide global mobile coverage, and the system was launched in late 1998 to its customers. Iridium provides global mobile coverage including the oceans, airways, and polar regions, and there is excellent satellite visibility and coverage at both the North and South Poles. The existing fixed line and mobile telecom systems had limited coverage in remote areas, and so the concept of global coverage as proposed by Iridium was potentially very useful.

Iridium was implemented by a constellation of 66 satellites. The original design required 77 satellites, and so the name *Iridium* was chosen since its atomic number in the periodic table is 77. However, the later design required just 66 satellites, and so *Dysprosium* may have been a more appropriate name. The satellites are in low Earth orbit hundreds of miles above the Earth, and communication between the satellites is via inter-satellite links. Each satellite can have four inter-satellite links, and each satellite contains seven Motorola Power PC 603E processors running at 200 MHz. These machines are used for satellite communication and control, and the satellites have an onboard fault detection system that allows for rapid and safe mitigation of system issues.

© Springer Nature Switzerland AG 2018
G. O'Regan, *The Innovation in Computing Companion*,
https://doi.org/10.1007/978-3-030-02619-6_34

Iridium routes phone calls through space, and there are several Earth stations. As satellites leave the area of an Earth base station, the routing tables change, and frames are forwarded to the next satellite just coming into view of the Earth base station.

Iridium is a large commercial satellite constellation, and it is especially suited for industries such as maritime, aviation, government, and the military. Motorola was the prime contractor for its design and development, and the satellites were produced at a cost of $5 million each ($40 million each including launch costs). Motorola engineers could make a satellite in the phenomenal time of 2–3 weeks (Fig. 34.1), and Motorola used launch vehicles from 3 companies (McDonnell Douglas, Long March IIC, and Proton-K) in 3 different countries (the USA, China, and Russia).

The first Iridium call was made by Al Gore in late 1998. However, despite being an engineering triumph, Iridium was a commercial failure, and it went bankrupt in late 1999 due to insufficient demand for its services. Its revenue was insufficient to service the debt associated with building the constellation of satellites, and it had needed a million subscribers to break even. The cost of making an Iridium call was very expensive compared to the existing cellular providers, and the cost of its handsets was much higher and more cumbersome to use than existing mobile phones. This meant that there was very little demand for its services, and the key reasons for failure include:

– Insufficient demand for its services (10,000 subscribers)
– High cost of its service ($5 per minute for a call)
– Cost of its mobile handsets ($3000 per handset)
– Bulky mobile handsets
– Competition from existing mobile phone networks
– Management failures

However, the Iridium satellites remained in orbit, and the service was re-established in 2001 by the newly founded Iridium Satellite LLC. The new business and pricing model required just 60,000 subscribers to break even, and it was targeted

Fig. 34.1 Replica of a first-generation Iridium satellite. Creative Commons

to a niche market of journalists, explorers, and the military who require reliable service in remote parts of the globe. Today, Iridium has over half a million customers, and it is used extensively by the US Department of Defense.

Iridium was designed in the late 1980s, and so it is designed primarily for voice rather than data. This means that it lacks the sophistication of modern mobile phone networks, and it is not as attractive to users. However, it provides service in remote parts of the world, which is very useful

Iridium Next is the second generation of Iridium and includes voice and data transmission. The network is being built by Thales Alenia, and the existing Iridium satellites are expected to remain in use until Iridium Next is fully operational in the 2020s. The new satellites are being launched by SpaceX, which is an American aerospace manufacturer and space transport company, and the Falcon launch rockets will launch 70 Iridium Next satellites.

Chapter 35
Java Programming Language

The rise of the Internet and the World Wide Web led to fundamental changes in the computing field. The previous paradigm was dominated by stand-alone personal computers that could be connected to file servers, whereas today nearly all computers are connected to the Internet. The Internet and World Wide Web were initially used to share files and information, whereas today the World Wide Web is a vast distributed computing space. *Java has played a key role in transforming the Internet*, and it has fundamentally changed the way in which people program. *It supports application programming in a distributed computing environment.*

Java is a collection of software packages and specifications created by Sun Microsystems in the mid-1990s. The Java platform and language began as an internal project at Sun in 1990, with the goal of developing a platform-independent language that could be for embedded software development. The plan was that the Java language could be used to develop embedded software for consumer devices such as toasters and microwave ovens. The project team included James Gosling and Patrick Naughton, and their mission was to create a cross-platform language that would run on a variety of CPUs in different environments and could be easily ported to many types of devices.

James Gosling is a Canadian computer scientist and is regarded as the father of the Java programming language (Fig. 35.1). Gosling and others at Sun Microsystems designed the language and implemented the original compiler and virtual machine. The language was originally called "Oak," but it was renamed to "Java" in 1995. It is a popular language for developing application software and may be deployed in a cross-platform computing environment.

The team initially considered using the C++ programming language, but they rejected it as it used too much memory in the development of embedded systems. Further, they believed that its complexity led to developer errors, and they also had concerns about its garbage collection and the portability of the language.

The invention of the World Wide Web in the late 1980s had a major impact on the development of Java. Portability was a key requirement for the Web, and it was clear to Gosling and the others that the problem of portability encountered when creating

© Springer Nature Switzerland AG 2018

G. O'Regan, *The Innovation in Computing Companion*,

https://doi.org/10.1007/978-3-030-02619-6_35

Fig. 35.1 James Gosling

software for embedded controllers is also encountered when creating software for the Internet. The latter is a distributed system consisting of many types of computers, operating systems, and central processing units. This led to a change in the focus of Java from embedded software to Internet programming.

Gosling initially tried to modify and extend C++, but he abandoned it in favor of a new language, which was initially called "Oak" and renamed to "Java" in 1995.[1] The syntax of the language is adapted from the C programming language, and the object-oriented features are adapted from C++. The team demonstrated part of the new platform in mid-1992, including the Oak language, the libraries, and the hardware. They piloted the new platform in the development of a personal digital assistant (PDA) later that year, and in mid-1994 the platform was targeted toward the World Wide Web. The first public release of the language was at the Sun World conference in mid-1995. Sun established the JavaSoft group to continue its development.

Java has had a major impact on the Internet, and it has simplified web programming. It introduced a fundamentally new network program called the *Java applet*, which is a special type of Java program that is transferred over the Internet and automatically executed by the web browser. An applet is downloaded on demand without further user interaction. That is, if a user clicks a link that contains a Java applet, then the applet will be automatically downloaded and run by the browser.

---

[1] The word "Oak" is a trademark of Oak Technology, and so the language was renamed to Java.

An applet is a dynamic self-executing program that is initiated by the server and runs on the client machine. The creation of the applet fundamentally changed Internet programming, as it brought the world of dynamic self-executing programs into cyberspace. Java includes security features to prevent the applet from acting as malicious software by restricting the applet to the Java execution environment and preventing it from accessing other parts of the computer. This allows applets to be downloaded safely as well as providing confidence that no harm will be done and that security will not be breached.

Portability is a key concern for the Internet given the diverse range of computers and environments connected to it. The same code must work on all computers, as it is not practical to have different versions of the applet for all possible computers. This requires a way to generate portable executable code, and Java solves the portability problem in that the output of a Java compiler is *bytecode* and not executable code. Bytecode is an optimized set of instructions designed to be run by the *Java Virtual Machine* (JVM). The JVM is the Java run-time system, and the original JVM was designed as an interpreter for bytecode.

The translation of Java into bytecode solves the portability problems for web-based programs, and it allows the program to be executed in many different environments. The JVM needs to be implemented for each environment, and once this is done, any Java program may run on it. The JVM will vary from platform to platform, but it understands bytecodes and executes the program in its run-time environment. The fact that the program is run by the JVM helps to make it secure, since the JVM is in control, and it can prevent the program from generating side effects outside the system.

An interpreted program is generally slower than compiled executable code. However, for Java the difference is not as large since the bytecode has been highly optimized. Further, a *just-in-time* (JIT) compiler may be part of the JVM, and it can compile selected portions of the bytecode into executable code in real time. A JIT compiler compiles code as it is needed during execution.

There have been many changes to Java since the original Java Developer Kit (JDK) was released in early 1996. There have been many additions of classes and packages to the standard library, and the evolution of Java is controlled through the Java Community Process (JCP), which allows users to propose additions and changes to the Java platform. The Java run-time environment is on millions of personal computers, and Java ME (Java Platform Micro Edition) has become popular in mobile devices. Google's Android operating system uses the Java language.

35.1   Java Virtual Machine

Java is a general-purpose object-oriented programming language, and the following is an example of the Hello World program written in Java.

```
class HelloWorld
{
     public static void main (String args[])
     {
          System.out.println ("Hello World!");
     }
}
```

Java was designed with portability in mind, and the objective is to create a computing environment that allows a program to be written once and executed anywhere. *Platform independence* is achieved by compiling the Java code into Java *bytecode, where the bytecode* is an optimized set of instructions.

The bytecode is then run on a *Java Virtual Machine* (JVM) that interprets and executes the bytecode. The JVM is specific to the native code of the host hardware. Java also provides automatic garbage collection, which protects programmers who forget to deallocate memory (thereby causing memory leaks).

A *native compiler* compiles the source code into the equivalent executable code. The Java compiler compiles the source code to the object code of a virtual machine, and the translator module of the virtual machine translates each bytecode of the virtual machine to the corresponding native machine instruction. That is, the virtual machine translates each generalized machine instruction into a specific machine instruction (or instructions) that may then be executed by the processor on the target computer.

Most computer languages such as C require a separate compiler for each computer platform (i.e., computer and operating system). However, a language such as Java comes with a virtual machine for each platform. This allows the source code statements in these programs to be compiled just once, and they will then run on any platform. There are many texts available on the Java programming language, and for more detailed information, see, for example, (Arnold et al. 2013; Gosling et al. 2014).

Chapter 36
LEO Computers

The LEO I computer was one of the earliest business computers, and it was developed by J. Lyons and Co. It was modeled on the Cambridge EDSAC computer, which was designed by Wilkes and others at the University of Cambridge. Lyons partially funded the development of EDSAC, and Lyons set up a project team led by John Pinkerton to develop its own computer that would be suitable for business applications. Wilkes provided training for Lyon's engineers, and the LEO computer ran its first program in late 1951.

J. Lyons and Co. was a well-known British conglomerate that was founded in the late nineteenth century. It was an innovative and forward-thinking company, and it was committed to finding ways to continuously improve to serve its customers better. It sent two of its executives to the United States shortly after the Second World War to evaluate new methods to improve business processes. The executives came across the early computers that had been developed in the United States, including the ENIAC computer that had been developed by John Mauchly and others. They recognized the potential of these early machines for business data processing.

They also became aware during their visit to the United States that Maurice Wilkes and others at the University of Cambridge in England were working on the design of a computer based on the ideas detailed in von Neumann's report (see Chap. 54). On their return to England, they visited Wilkes at the University of Cambridge, who was working on the design of the EDSAC computer. They were impressed by his ideas and technical knowledge and recognized the potential of the planned EDSAC computer.

They prepared a report for Lyon's board recommending that a computer designed for data processing should be the next step in improving business processes and that Lyons should develop or acquire a computer to meet its business needs. Lyons and the University of Cambridge entered a collaboration arrangement where Lyons agreed to help fund the completion of EDSAC and the University of Cambridge agreed to help Lyons to build its own computer.

This led to the Lyons Electronic Office (LEO) I computer, which was based on the Cambridge EDSAC but adapted to business data processing. The LEO I was

© Springer Nature Switzerland AG 2018
G. O'Regan, *The Innovation in Computing Companion*,
https://doi.org/10.1007/978-3-030-02619-6_36

initially used for valuation jobs, but this was later extended to other business applications. Lyons was one of the pioneers of IT outsourcing as it performed payroll calculations for several companies in the United Kingdom.

Lyons recognized that more and more companies would require computing power, and they saw a business opportunity. They decided to set up a subsidiary company (LEO Computers Ltd.) to focus on computers for commercial applications. The company was founded in London in 1954, and it was one of the earliest British computer companies. It designed and manufactured the LEO I, LEO II, and LEO III business computers. Several British companies purchased the LEO II computer, and the LEO III was sold to customers in the United Kingdom and overseas.

The LEO computers were mainly used for business applications such as valuation, payroll, and inventory. LEO Computers Ltd. merged with English Electric in 1963 to become English Electric LEO, and it later became English Electric LEO Marconi (EELM). This company became part of the International Computers Ltd. (ICL) in 1968. For more detailed information on J. Lyons and Co. and the LEO computers, see (Ferry 2003).

36.1 LEO I Computer

Lyons and the University of Cambridge entered a collaboration arrangement where Lyons agreed to help fund the completion of EDSAC, and the University of Cambridge agreed to help Lyons to develop their own computer, which was called the LEO I computer (Fig. 36.1). This machine was based on EDSAC but adapted to business data processing.

The Electronic Delay Storage Automatic Calculator (EDSAC) was completed and ran its first program in 1949, and the LEO I computer was completed and ran its first program in late 1951. Lyons developed several applications for the LEO computer, and soon the LEO I computer was used to process business applications (e.g., payroll) for other companies.

LEO I's clock speed was 500 kHz with most instructions taking 1.5 milliseconds to complete. The machine was linked to fast paper tape readers and fast punched card readers and punches. It had 2048 35-bit words of memory (i.e., $2048 \times 35/8 = 8.75$ Kb). It had 6000 valves in 21 racks, and it consumed 30 kW of power.

The LEO I was initially used for valuation jobs, but this was later extended to payroll, inventory, and other applications. One of the early applications developed by Lyons was an early version of an integrated management information system to manage its business. Lyons was also one of the pioneers of IT outsourcing in that it performed payroll calculations for several companies in the UK.

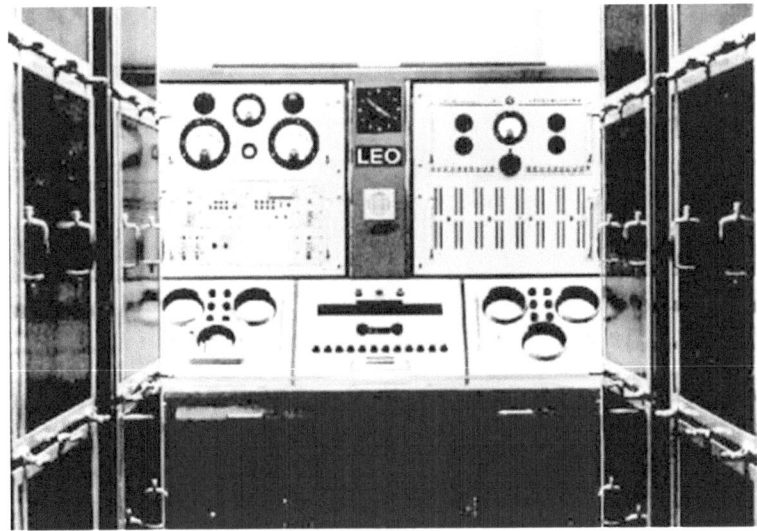

Fig. 36.1 LEO I computer. Courtesy of LEO Computer Society

The UK Met Office used the LEO I computer in an early attempt at using a computer for weather forecasting in the early 1950s. The weather prediction model was solved on the LEO I computer, and the first predictions were made in 1954. The Met Office later used the Manchester Mark I (see Chap. 37) and more powerful computers for weather forecasting.

36.2 LEO II and LEO III Computers

The LEO II computer was an improved version of the LEO I, and it was released in 1957. Several models of the LEO II computers were delivered, and the machine had 4.875 Kb of memory. The LEO III was introduced in 1961.

LEO II computers were installed in several offices in Britain including the Will Tobacco Co., Ford Motor Company, British Oxygen Company, and the Ministry of Pensions in Newcastle.

The LEO III (Fig. 36.2) was a transistorized computer, and it was installed in Customs and Excise, Inland Revenue, and the Post Office. It was also sold in Australia, South Africa, and Czechoslovakia.

The final machines developed were the LEO 360 and LEO 326, and the final LEO in use (the LEO 326) shut down for the last time in 1981.

Fig. 36.2 LEO III computer. Courtesy of LEO Computer Society

Chapter 37
Manchester Baby and Mark 1 Computers

The Manchester Small-Scale Experimental Computer (better known by its nickname "Baby") was developed at the University of Manchester (Fig. 37.1). It was the *first stored-program computer*, and it was designed and built at the University of Manchester in England by Frederic Williams, Tom Kilburn, and others.

Kilburn was assisted by Geoff Tootill in the design and construction of the prototype machine, and the machine demonstrated the reliability of the Williams tube. It was the first stored-program computer: i.e., the instructions to be executed were loaded into memory rather than rewiring the computer to run a new program. That is, all that was required was to enter a new program into the computer memory, rather than rewiring the machine to solve a new problem (as required with ENIAC and discussed in Chap. 23). Kilburn wrote and executed the first stored program, which was a short 17-line program to compute the highest proper divisor of 2^{18}. The program was written and executed in 1948, and it took 52 min to compute the correct answer (131,072).

The prototype machine demonstrated the feasibility and potential of a stored-program computer. Its memory consisted of 32 32-bit words, and it took 1.2 milliseconds to execute one instruction, i.e., 0.00083 MIPS (million instructions per second). Today's computers are rated at speeds such as 1000 MIPS. The team in Manchester developed the machine further, and in 1949, the Manchester Mark 1 was available.

37.1 Manchester Mark 1

The Manchester Automatic Digital Computer (MADC), also known as the Manchester Mark 1, was developed at the University of Manchester (Fig. 37.2). It was one of the earliest stored program computers, and it was the successor to the Manchester "Baby" computer. It was designed and built by Williams, Kilburn, and others.

© Springer Nature Switzerland AG 2018
G. O'Regan, *The Innovation in Computing Companion*,
https://doi.org/10.1007/978-3-030-02619-6_37

Fig. 37.1 Replica of the Manchester Baby. Courtesy of Tommy Thomas

Each word could hold one 40-bit number or two 20-bit instructions. The main memory consisted of two pages (i.e., two Williams tubes with each holding 32 × 40-bit words or 1280 bits). The secondary backup storage was a magnetic drum consisting of 32 pages (this was updated to 128 pages in the final specification). Each track consisted of two pages (2560 bits). One revolution of the drum took 30 milliseconds, and this allowed the 2560 bits to be transferred to the main memory.

It contained 4050 vacuum tubes and had a power consumption of 25,000 watts. The standard instruction cycle was 1.8 milliseconds, but multiplication was much slower. The machine had 26 defined instructions, and the program instructions were entered in binary format, as assembly languages and assemblers were not yet available.

It had no operating system, and its only systems software were some basic routines for input and output. Its peripheral devices included a teleprinter and a five-hole paper tape reader and punch.

A display terminal used with the Manchester Mark 1 computer mirrored what was happening within the Williams tube. A metal detector plate placed close to the surface of the tube could detect changes in electrical charges. The metal plate obscured a clear view of the tube, but the technicians could monitor the tubes used with a video screen. Each dot on the screen represented a dot on the tube's surface, and the dots on the tube's surface worked as capacitors that were either charged and bright or uncharged and dark. The information translated into binary code (0 for dark, 1 for bright) became a way to program the computer.

Fig. 37.2 The Manchester Mark 1 Computer. Courtesy of the University of Manchester

37.2 Williams Tube

Frederick Williams and Tom Kilburn experimented on the use of cathode-ray tubes as an information storage device. They developed what became known as the *Williams-Kilburn tube*, which was the first form of random-access memory. This digital storage device was used successfully on several early computers.

The Williams tube was a cathode-ray tube used to store binary data, and each Williams tube could store 512–1024 bits of data. It was the first random-access digital storage device, and it remained popular in the computer data storage field for several years until it was outdated by core memory in the mid-1950s. It provided the first large amount of random-access memory (RAM), and it was used as a key component in the first stored-program computer.

Williams had succeeded in storing 1 bit of information on a cathode-ray tube, and Kilburn began working with him in the mid-1940s to improve its digital storage ability. Kilburn devised an improved method of storing bits, which increased the storage capacity. They were now ready to build a computer to test the reliability of the memory in the Williams tube, and this led to the development of the small-scale experimental computer (SSEC), which was the *first stored-program digital electronic computer* (popularly known as the Manchester "Baby") in 1948.

37.3   Ferranti Mark 1

Ferranti Ltd. (a British company) and the University of Manchester collaborated to build one of the first commercial general-purpose electronic computers. This was the Ferranti Mark 1, and it was basically an improved version of the Manchester Mark 1. The first machine off the production line was delivered to the University of Manchester in 1951 and shortly before the release of the UNIVAC 1 electronic computer in the United States.

The Ferranti Mark 1's instruction set included a "hoot command" which allowed auditory sounds to be produced, and it also allowed variations in pitch. Christopher Strachey (who later did important work on the semantics of programming languages) programmed the Ferranti Mark 1 to play tunes such as "God Save the King,"[1] and the Ferranti Mark 1 was one of the earliest computers to play music. The parents of Tim Berners-Lee (the inventor of the World Wide Web) both worked on the Ferranti Mark 1.

The main improvements of the Ferranti Mark 1 over the Manchester Mark 1 were improvements in size of primary and secondary storage, a faster multiplier, and additional instructions.

It had eight pages of random-access memory (i.e., 8 Williams tubes each with a storage capacity of 64 20-bit words or 1280 bits). The secondary storage was provided by a 512-page magnetic drum that stored 2 pages per track, and its revolution time was 30 milliseconds.

It used a 20-bit word stored as a single line of dots on the Williams tube display, with each tube storing a total of 64 lines of dots (or 64 words). Instructions were stored in a single word, while numbers were stored in two words.

The accumulator was 80 bits, and it could also be addressed as two 40-bit words. There were about 50 instructions, and the standard instruction time was 1.2 milliseconds. Multiplication could be completed in 2.16 milliseconds. There were 4050 vacuum tubes employed.

---

[1] Queen Elizabeth II became the reigning monarch after the death of her father, King George VI, in 1952.

Chapter 38
Microprocessor

The invention of the microprocessor (initially called microcomputer) in 1971 was a revolution in computing, with the power of a computer now available on a tiny microprocessor chip. The microprocessor is essentially a computer on a chip, and its invention made handheld calculators and personal computers (PCs) possible. Intel's microprocessors are used on most personal computers and laptops around the world.

Intel introduced the world's first microprocessor, the *Intel 4004*, in 1971, and the company later became the industry leader in the microprocessor field. Although the Intel 4004 was initially developed as an enhancement to allow users to add more memory to their units, it soon became clear that the microprocessor had applications to everything from calculators to cash registers and traffic lights.

The invention of the microprocessor happened by accident rather than design, when Nippon Calculating Machine Corporation (later known as Busicom) requested Intel to design a set of integrated circuits for its new family of high-performance programmable calculators. At that time, it was standard practice to custom design all logic chips for each customer's product, and this clearly limited the applicability of a logic chip to a specialized domain.

Busicom's proposed design required 12 integrated circuits, and Ted Hoff, an Intel engineer, studied Busicom's design and proposed a more elegant solution. Hoff's design required just four integrated circuits and included a chip that was a general-purpose logic device (microprocessor) that derived its application instructions from the semiconductor memory. Busicom accepted his proposed design, and Intel engineers implemented it.

Hoff's 4004 microprocessor design included a central processing unit (CPU) on one chip. It contained 2300 transistors on a one-eighth by one-sixth-inch chip surrounded by three ICs containing ROM, shift registers, input/output ports, and RAM.

Busicom had exclusive rights to the design and components, but following discussion and negotiations, Busicom agreed to give up its exclusive rights to the chips. Intel shortly afterward announced the availability of the first microprocessor, the Intel 4004 (Fig. 38.1).

© Springer Nature Switzerland AG 2018

G. O'Regan, *The Innovation in Computing Companion*,
https://doi.org/10.1007/978-3-030-02619-6_38

Fig. 38.1 Intel 4004
microprocessor

This was the world's first microprocessor, and although it was initially developed as an enhancement that allowed users to add more memory to their units, it soon became clear that the microprocessor could be applied to many other areas. It was launched in late 1971, and it could execute 60,000 operations per second. That is, this tiny chip had an equivalent computing power as the large ENIAC computer that used 18,000 vacuum tubes and took up the space of an entire room (see Chap. 23).

The Intel 4004 sold for $200 and for the first-time affordable computing power was available to designers of all types of products. The introduction of the microprocessor was a revolution in computing, and its invention had applications to everything from traffic lights to medical instruments and to the development of home and personal computers.

Intel was founded by Robert Noyce and Gordon Moore, and its initial focus was on semiconductor memory products and to create large-scale integrated (LSI) semiconductor memory. Today, it is a large semiconductor manufacturer that dominates the microprocessor market for PCs and laptops. It has a broad product line including motherboards, flash memory, switches, and routers, and through continuous innovation, it has introduced more and more powerful microprocessors since its invention of the Intel 4004.

Gary Kildall was one of the early people to recognize the potential of the microprocessor as a computer, and he began writing experimental programs for the Intel 4004 in the early 1970s. He worked as a consultant with Intel on the later 8008 and 8080 microprocessors.

Kildall developed the first high-level programming language for a microprocessor (PL/M) in 1973, which enabled programmers to write applications for microprocessors. He developed the CP/M operating system (Control Program for Microcomputers) in the same year (see Chap. 42), which allowed the Intel 8080 microprocessor to control a floppy disk drive allowing files to be read and written to and from an eight-inch floppy disk. CP/M made it possible for computer hobbyists and companies to build the first home computers.

Kildall made CP/M hardware independent by creating a separate module called the BIOS (basic input/output system). He added several utilities such as an editor, debugger, and assembler, and by 1977, several manufactures were including CP/M with their systems. He set up Digital Research Inc. (DRI) in 1976 to develop, market, and sell the CP/M operating system.

38.1 Early Microprocessors

Intel has developed more and more powerful microprocessors since its invention of the Intel 4004 in 1971. The 8-bit Intel 8008 was launched in 1972, and this reasonably successful product led to the 8-bit Intel 8080 microprocessor, which was released in 1974. The Intel 8080 had 45,000 transistors, and it was the first general-purpose microprocessor. It was sold for $360: i.e., a whole computer on one chip was sold for $360, while conventional computers were sold for thousands of dollars. The Intel 8080 soon became the industry standard, and Intel became the industry leader in the 8-bit market. The 8080 played an important role in starting personal computer development, as it attracted the interest of developers and engineers. The 8-bit Intel 8085 was introduced in 1976.

Motorola introduced its first microprocessor, the 8-bit 6800 microprocessors (Fig. 38.2) in 1974, and this microprocessor was used in automotive, computing, and video games. It contained over 4000 transistors. It competed against the Intel 8080 microprocessor, and it was used in some early home computer kits.

National Semiconductor introduced its 16-bit IMP-16 in 1973 and an 8-bit version, the IMP-8, in 1974. Texas Instruments introduced its first single-chip microprocessor, the PACE, in 1974, and it introduced its first 16-bit microprocessor, the TMS 9900, in 1976. MOS Technology introduced its 8-bit 6502 in 1975, and Zilog introduced its Z80 in 1976.

The 16-bit Intel 8086 was introduced in 1978, but it soon faced competition from Motorola, which introduced its 16-/32-bit 68000 microprocessor in 1979. The Intel 8088 is an 8-bit variant of the 8086, and it was introduced in 1979. The Motorola 68000 was a hybrid 16-/32-bit microprocessor that had a 16-bit data bus, but it could perform 32-bit calculations internally. It was used on various Apple Macintosh computers, the Atari ST, and the Commodore Amiga.

The first single-chip 32-bit microprocessor was AT&T Bell Labs BELLMAC-32A, which was introduced in 1982. Motorola introduced its 32-bit 68020 microprocessor in 1984, and this microprocessor contained 200,000 transistors on a three-eighths-inch square chip.

IBM considered several microprocessors for its IBM PC including the IBM 801 processor, the Motorola 68000 microprocessor, and the Intel 8088 microprocessor. It chose the Intel 8088 chip (which was cheaper than the 16-bit Intel 8086), and it took a 20% stake in Intel leading to strong ties between both companies.

Intel has developed more and more powerful microprocessors over the years. It introduced the 16-bit 80286 microprocessor in 1982; the 32-bit 80386 microprocessor

Fig. 38.2 Motorola 6800
microprocessor

with 275,000 transistors appeared in 1985; the 80486 microprocessor containing over 1 million transistors was released in 1989; the Pentium I processor containing 3 million transistors was released in 1993; the Pentium II with over 7 million transistors appeared in 1997; and the Pentium 4 with 42 million transistors was released in 2000.

The 32-bit 80486 microprocessor appeared in 1989, and Business Week described it as "a verifiable mainframe on a chip." It had 1.2 million transistors and the first built-in math coprocessor.

Today, Intel's microprocessors are used on most personal computers and laptops around the world, and the contract to supply the 8-bit Intel 8088 microprocessor was a major turning point for the company. Intel had been focused more on the sale of dynamic random-access memory chips, with sales of microprocessors in thousands or in tens of thousands. However, sales of microprocessors rocketed following the introduction of the IBM PC, and soon sales were in tens of millions of units. For a more detailed account of Intel, see (Malone 2014).

The introduction of the IBM PC was a revolution in computing, and there are hundreds of millions of computers in use around the world today. The microprocessor placed computing power in the hands of ordinary users, and today's personal computers are more powerful than the mainframes that were used to send man to the moon. The cost of computing processing power has fallen exponentially since the introduction of the first microprocessor, and the continuous innovation of semiconductor companies has squeezed more and more transistors onto a chip leading to more and more powerful microprocessors and personal computers.

Chapter 39
Mobile Phone

The invention of the telephone by Graham Bell in the late nineteenth century was a revolution in human communication, as it allowed people in different geographic locations to communicate instantaneously rather than meeting face-to-face. However, the key restriction of the telephone was that the actual physical location of the person to be contacted was required prior to communication, as otherwise communication could not take place: i.e., *communication was between places rather than people.*

The origin of the mobile phone revolution dates back to work done on radiotechnology in the 1940s. Bell Labs had proposed the idea of a cellular communication system back in 1947, and it was eventually brought to fruition by researchers at Bell Labs and Motorola. Bell Labs constructed and operated a prototype cellular system in Chicago in the late 1970s and performed public trials in 1979. Motorola commenced a second US cellular system test in the Washington/Baltimore area, and the first commercial systems began in the United States in 1983.

DynaTAC (Dynamic Adaptive Total Area Coverage) used cellular radiotechnology to link people and not places. Motorola was the first company to incorporate the technology into a portable device designed for use outside of an automobile, and it spent $100 million on the development of cellular technology. Martin Cooper (Fig. 39.1) led the team at Motorola that developed the DynaTAC 8000X, and he made the first mobile phone call on a prototype DynaTAC phone to Joel Engels, the head of research at Bell Labs, in April 1973.

Commercial cellular services commenced in North America in 1983, and the world's first commercial mobile phone went on sale the same year. This was the Motorola DynaTAC 8000X, and it was popularly known as the *brick* due to its size and shape. It weighed 28 ounces (almost 2 lbs); it was 13.5″ (over a foot) in length and 3.5″ in width. It had a LED display and could store 30 numbers. It had a talk time of 30 min and 8 h of standby; and it took over 10 hours to recharge.

The cost of the Motorola DynaTAC 8000X was $3995, and it was too expensive for most people apart from wealthy consumers. Today, mobile phones are ubiquitous,

© Springer Nature Switzerland AG 2018
G. O'Regan, *The Innovation in Computing Companion*,
https://doi.org/10.1007/978-3-030-02619-6_39

Fig. 39.1 Martin Cooper
reenacts DynaTAC call

and there are more mobile phone users than fixed line users. The cost of a mobile phone today is often less than $100, and it weighs as little as 3 ounces.

The first-generation mobile phone system introduced into North America in the early 1980s used the 800 MHz cellular band. It had a frequency range between 800 and 900 MHz. Each service provider could use half of the 824–849 MHz range for receiving signals from cellular phones and half the 869–894 MHz range for transmitting to cellular phones. The bands were divided into 30 kHz sub-bands called channels, and a separate frequency (or channel) was used for each conversation. The division of the spectrum into sub-band channels is achieved by using frequency division multiple access (FDMA).

This first-generation system allowed voice communication only, and it was susceptible to static and noise. Further, it had no protection from eavesdropping using a scanner.

The AXE system (discussed in Chap. 11) provided the foundation for Ericsson's growth in mobile telephony. The flexible modular design of AXE allowed new functionality to be added, and by changing a module, AXE could be reconfigured to handle mobile telephone calls. This allowed Ericsson to design the first mobile telephone exchange (MTX) by replacing the subsystem for fixed subscribers with a new subsystem for mobile subscribers. The MTX switch was developed in the late 1970s/early 1980s and was a key part of the Nordic Mobile Telephone (NMT) system which would be used in all Nordic countries.

Ericsson was awarded a large Saudi Arabian contract to deliver a fixed line and mobile system, and it was agreed that the NMT standard would be used and that

Ericsson would supply the entire system. The Saudi mobile phone network became operational from 1981, and Ericsson provided base stations, radio towers, and switches. Ericsson had now acquired cell-planning experience, and it was awarded the contract to develop the entire mobile telephone network in the Netherlands. Ericsson was now a total systems supplier in mobile telephony, and it provided the entire infrastructure such as switches and base stations. Today, its base stations range from small picocells to large macrocells.

The second generation (2G) of mobile technology was a significant improvement on the existing analog technology. This digital, cellular technology encrypted telephone conversations and provided data services such as text and picture messages. The main second-generation technologies were the GSM standard developed by the European Telecommunications Standards Institute (ETSI) and CDMA developed in the United States. The first GSM call was made by the Finnish prime minister in Finland in 1991, and the first short message service (SMS) or text message was sent in 1992.

The subscriber identity module (SIM) card was a new feature in GSM, and this is a detachable smart card that contains the user's subscription information and phone book. The SIM card may be used in other GSM phones, and this is useful when the user purchases a replacement phone. GSM provides an increased level of security, with communication between the subscriber and base station encrypted.

GSM networks evolved into GPRS (2.5 G), which became available in 2000. Third- and fourth-generation (3G and 4G) mobiles provide mobile broadband multimedia communication. Mobile phone technology has transformed the earlier paradigm of *communication between places* to that of *communication between people*.

Motorola dominated the analog mobile phone market, but it was slow in adapting to the GSM standard, and it paid a heavy price with a loss of market share to Nokia and Ericsson. It was very slow to see the potential of a mobile phone as a fashion device,[1] and it was too slow in adapting to smartphones. It paid the ultimate price, and today it is a shadow of its former self.

39.1 Development of Mobile Phone Standards

Bell Labs played an important role (with Motorola) in the development of the analog mobile phone system in the United States. It developed a system in the mid-1940s that allowed mobile users to place and receive calls from automobiles, and Motorola developed mobile phones for automobiles. However, these phones were large and bulky and they consumed a lot of power. A user needed to keep the automobile's engine running to make or receive a call.

Bell Labs first proposed the idea of a cellular system back in the late 1940s, and it proposed hexagonal rings for mobile communication. Large geographical areas

---

[1] The attitude of Motorola at the time seemed to be like that of Henry Ford: i.e., they can have whatever color they like as long as it is black.

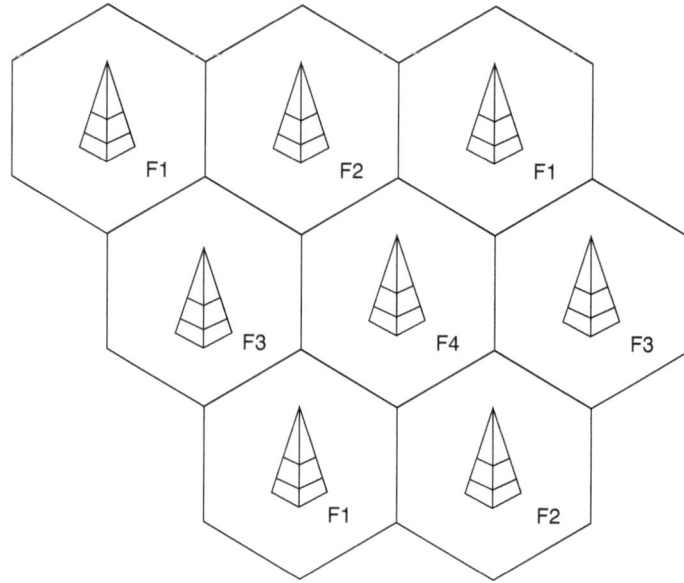

Fig. 39.2 Frequency reuse in cellular networks

were divided into cells, where each cell had its own base station and channels. The available frequencies could be used in parallel in different cells without disturbing each other (Fig. 39.2). Mobile telephone could now, in theory, handle many subscribers. However, it was not until the late 1960s that Bell Labs prepared a detailed plan for implementing the cellular system.

Bell Labs developed the Advanced Mobile Phone System (AMPS) standard from 1968 to 1983. Motorola and other telecommunication companies designed and built phones for this cellular system. The AMPS standard uses separate frequencies (or channels) for each conversation and requires considerable bandwidth for many users. The signals from a transmitter cover an area called a cell, and as a user moves from one cell into a new cell, a handover to the new cell takes place without any noticeable change to the user. The signals in the adjacent cell are sent and received on different channels to the existing cell's signals, and so there is no interference.

The Total Access Communication (TACS) and Extended TACS (ETACS) system are variants of AMPS that were employed in the United Kingdom and Europe. These analog standards employed separate frequencies (or channels) for each conversation using frequency division multiple access (FDMA). However, the analog system suffered from static and noise, and there was no protection from eavesdropping using a scanner.

Ericsson and Motorola became the leaders in the first generation of mobile, and the extent of Ericsson's leadership was clear when its proposed design for digital mobile radio transmission was selected as the US Standard for Cellular Communications over entries from Motorola and AT&T in 1989.

The AMPS standard is a first-generation technology, and mobile technology has evolved to second-generation digital Global System for Mobile Communication (GSM) and Code Division Multiple Access (CDMA) technologies; to General Packet Radio Service (GPRS); to third-generation mobile, including 3G and WCDMA; and to fourth-generation and fifth-generation mobile (4G and 5G).

Chapter 40
Mouse

The computer mouse was invented by Douglas Engelbart of the Augmentation Research Centre (ARC) at the Stanford Research Institute (SRI) in the mid-1960s. It consisted of a wooden shell, a circuit board, and two metal wheels that came into contact with the surface that it was being used on (Fig. 40.1). Engelbart had been investigating ways for individuals to improve their capability in solving complex problems, and the mouse was part of ARC's online system (NLS).

Engelbart envisaged problem-solvers using computer-aided work stations using some sort of device to move a cursor around a screen. Engelbart and Bill English developed the first prototype of the mouse in 1964, and it worked on an early windows graphical user interface. They christened the device *mouse* as the early prototypes had a cord attached to the rear part of the device that looked like a tail and resembled an actual mouse.

The 1967 patent application described the mouse as an X-Y position indicator for a display system. It was publicly demonstrated at a famous computer conference in 1968, where Engelbart and a group of 17 other researchers of the ARC group gave a public demonstration of their NLS system.

The public demonstration took place at the Fall Joint Computer Conference held in San Francisco, and the mouse was just one of several innovations presented by Engelbart on that day. The goal of the NLS system was to act as an instrument to help humans operate within the domain of complex information structures.

The demonstration included the mouse, hypertext,[1] a precursor to today's graphical user interfaces, and networked computers with shared screen collaboration involving two people at different sites communicating over a network with audio and video interface. The public demonstration introduced several fundamental computing concepts taken for granted today, and it later became known as *The Mother of all Demos*.

The mouse operates on the principle that the computer determines the distance and speed that the mouse has traveled and converts that information into coordinates

---

[1] Ted Nelson coined the term hypertext in the early 1960s.

© Springer Nature Switzerland AG 2018
G. O'Regan, *The Innovation in Computing Companion*,
https://doi.org/10.1007/978-3-030-02619-6_40

Fig. 40.1 SRI first mouse

Fig. 40.2 Two Macintosh
Plus mice. 1984

that it can plot on a display screen. The original mouse was used by Engelbart to navigate the NLS system, and this was also the first system to use hypertext.

Bill English moved to Xerox PARC in 1971, and the *ball mouse* was developed by English at PARC in 1972. It replaced the external wheels with a single ball that could rotate in any direction. The ball mouse became an important part of the graphical user interface of the Xerox Alto computer system, which was developed in Xerox and used at several universities. Xerox eventually commercialized a version of the Alto (the Xerox Star 8010) in 1981, and this was one of the earliest computers to be sold with a mouse.

The term *mouse* became an accepted term of the modern computer lexicon when it was introduced as a standard part of the Apple Macintosh in 1984 (Fig. 40.2).

Fig. 40.3 A Computer
mouse with two buttons
and a scroll wheel

Steve Jobs had visited PARC to see the Xerox Alto, and he licensed the technology from Xerox. The Apple Lisa and Macintosh both used a graphical user interface and a mouse, and Microsoft made the MS/DOS Microsoft Word program mouse compatible, and the first Microsoft mouse for the PC appeared in 1983. The mouse became pervasive after the release of the Apple Macintosh and later Atari and Amiga personal computers in the mid-1980s and the release of Microsoft Windows 3.0 in the early 1990s.

A mouse is a pointing device that detects two-dimensional motion relative to a surface, and it generally involves the motion of a pointer on a display. It is held in the user's hand and generally has one or more buttons and a scroll wheel (Fig. 40.3).

An optical mouse was invented in 1980, which eliminated the need for the use of the rolling ball. The latter often became dirty from rolling around leading to a negative impact on its performance. It was several years before optical mice became commercially viable, but today they have replaced ball-based mice and are supplied as a standard part of all new computers.

An optical mouse is an advanced computer-pointing device that uses a light-emitting diode (LED), an optical sensor, and digital signal processing (DSP) instead of the traditional ball mouse technology. Movement is detected by sensing changes in reflected light rather than the interpretation of a rolling sphere. Steve Kirsch of MIT and Mouse Systems Corporation and Richard Lyon of Xerox invented the first optical mouse independently of each other in 1980.

Chapter 41
MP3 Player and Digital Music

An MP3 player (digital audio player) is an electronic device that can play digital audio files (Fig. 41.1). The MP3 file format is widely used, and there are also several other digital audio formats. Kane Kramer invented the digital audio player in 1979, and he filed a patent for his invention (the IXI) in 1981. The patent was granted in the United Kingdom in 1985 and in the United States in 1987.

Kramer's IXI player was as big as a credit card and had a small LCD screen, navigation, and volume buttons. It could hold 8 MB of data with a capacity of 3.5 min of audio. Kramer's patent entered the public domain in 1988 due to funding issues in the renewal of his patent.

Fraunhofer, the prestigious German research institute, developed the audio compression technology that is known as MP3. Karlheinz Brandenburg led the team at Fraunhofer that developed the technology, and he is known as the father of MP3. It is a standard for audio compression that makes an audio music file smaller with little or no loss in sound quality. Fraunhofer obtained a patent for MP3 in 1989, and the data compression typically shrinks the original sound data down by a factor of 10.

Fraunhofer developed an early (but unsuccessful) MP3 player in the early 1990s. The first successful MP3 player (the AMP MP3 Playback Engine) was developed by Tomislav Uzelac, a Croatian programmer, at the University of Zagreb in 1997. Uzelac formed Advanced Multimedia Products (AMP) to exploit the technology, and two other University of Zagreb students ported AMP to the Windows operating system and called it Winamp. This became a free MP3 music player in 1998 and helped to generate major interest in MP3.

41.1 Digital Music

Vinyl records (also called phonograph disk records) were the dominant technology for music reproduction up to the late twentieth century. The music was inscribed in a spiral grove starting from the outer part of the record and ending toward the center

© Springer Nature Switzerland AG 2018
G. O'Regan, *The Innovation in Computing Companion*,
https://doi.org/10.1007/978-3-030-02619-6_41

Fig. 41.1 Creative MuVo
TX FM digital audio
player

of the disk. The records are described in terms of their diameter, rotation speed per minute, and time duration (e.g., the 12 inch LP at 33 1/4 rpm for long-playing albums and the 7 inch SP at 45 rpm for a single).

Digital music is the process of representing sound as numerical binary values, as distinct from the previous paradigm where analog media such as vinyl records or magnetic tape were used to represent and store sound. One popular source of digital music is the compact disk, where binary data is used to represent musical sounds (Fig. 41.2).

Philips and Sony invented compact disk technology, and its invention was a technological revolution in the music industry. Philips demonstrated a prototype compact disk audio player in 1979, and this showed the feasibility of using digital optical recording and playback to reproduce audio signals with superb quality. The basic principle employed in compact disk technology is that the laser reads the surface of the CD and decodes it as binary data.

The world's first compact disk (CD) was manufactured at a Philips factory near Hannover in Germany in 1982. It provided superior sound quality and scratch-free durability, and it was the beginning of the shift from analog to digital in music technology.

One of the earliest CDs to be manufactured was *The Visitors* by ABBA, which was produced at the Philips Polygram recording company. Philips introduced its CD 100 CD Player and the first CDs into Europe and the United States in early 1983. One of the earliest fully digitally recordings to be brought to market was Dire Straits *Brother in Arms* album, which was released on CD in 1985, and it sold over a million copies.

The compact disk rapidly became the medium of choice for the music industry. Over 250 billion CDs, 3.5 billion audio CD players, and 3 billion CD-ROM drives had been sold by 2009. The success of the CD led to the end of the vinyl LP era, with most music companies issuing new releases only on CDs from the early 1990s.

The compact disk has played a key role in the shift from analog music to digital, and it has laid the foundation for an extensive family of optical disks such as CD-ROM, CD-R, CD-RW, DVD, DVD-R, DVD-RW, and Blu-Ray. The capacity of

Fig. 41.2 Compact disk

a CD is typically up to 700 MB and provides up to 80 min of audio. The capacity of a Blu-Ray disk is between 25 GB and 100 GB.

The IEEE Milestone Award was given to Philips in 2009 in recognition of its contributions to the development of the compact disk, and in setting the technical standard in digital optical recording systems.

The Digital Versatile Disk (DVD) is a digital optical disk recording format developed by Philips, Sony, Toshiba, and Panasonic in the late 1990s. A DVD disk has similar physical dimensions as a CD, but it has a much larger capacity (typically 1 and 16GB). It is used for digital consumer video as a replacement for the VHS videotapes or for digital consumer audio.

Chapter 42
MS/DOS Operating System

The introduction of the IBM Personal Computer in 1981 was a major milestone in the computing field. IBM's traditional approach to product development was to develop a full proprietary solution. However, due to the aggressive time scales associated with the introduction of the IBM PC, it decided instead to outsource the development of the microprocessor to a small company called Intel and to outsource the development of the operating system to a small company called Microsoft. These decisions would later prove to be costly to IBM, as Microsoft and Intel reaped the benefits and later became technology giants.

The award of the contract to develop the operating system for the IBM PC to Microsoft was controversial. IBM had intended awarding the contract to Digital Research Inc. (DRI), which had introduced the CP/M operating system for several microprocessors. However, IBM and Digital Research were unable to agree terms for the licensing of CP/M for the IBM PC.

IBM outsourced the development of PC/DOS (PC Disk Operating System) to Microsoft,[1] which had developed a BASIC interpreter for several home computers. The terms of the deal were favorable to Microsoft, as it gave it the right to market and sell its own version of the operating system (i.e., MS/DOS) to other manufacturers as the operating system for their IBM compatible computers. The IBM compatible PCs later came to dominate the market, as the open architecture of the IBM PC allowed manufacturers to create clone machines compatible with the original IBM PC at a more competitive price.

Microsoft hired a consultant to port an existing CP/M operating system to the 8088 microprocessor, and it later became clear to Digital Research that the PC/DOS operating system had been derived from CP/M without permission. Digital Research considered suing Microsoft for copying all CP/M system calls in DOS 1.0, but the legal advice was unclear with respect to the existing copyright law as to what constituted infringement of copyright. There was no guarantee of success in any legal action against IBM, and considerable expense would be involved

---

[1] Microsoft was founded by Bill Gates and Paul Allen in 1975.

© Springer Nature Switzerland AG 2018
G. O'Regan, *The Innovation in Computing Companion*,
https://doi.org/10.1007/978-3-030-02619-6_42

Fig. 42.1 Gary Kildall

Kildall threatened IBM with legal action, and IBM agreed to offer both PC/DOS and DRI's CP/M-86 as the operating system for the IBM PC. However, as PC/DOS was priced at $60 and CP/M at $240, CP/M faded into obscurity.

Bloomberg Businessweek published an article with the title "The Man who could have been Bill Gates" in 2004 (Bloomberg Business Week Magazine 2004). It describes the background to the development of the operating system for the IBM PC and the failed negotiations between Digital Research Inc. and IBM.

Kildall (Fig. 42.1) became aware of early work taking place at Intel on micropro-cessors, and he recognized the potential of the microprocessor as a computer in its own right. He began writing experimental programs for the newly released Intel 4004 (discussed in Chap. 38), and he developed the first microprocessor disk operat-ing system and the first programming language and compiler for a microprocessor.

He developed the CP/M operating system (Control Program for Microcomputers) in 1973, which allowed the Intel 8080 microprocessor to control a floppy drive. CP/M combined the essential components of a computer at the microprocessor level, and it was the first disk operating system for a microcomputer. Kildall set up Digital Research Inc. (DRI) with his wife Dorothy in 1976 to develop, market, and sell the CP/M operating system. He made CP/M hardware independent and added several utilities such as an editor, debugger, and assembly.

IBM approached Digital Research Inc. in 1980 with the intention of licensing their CP/M operating system for the new IBM personal computer. However, their negotiations were unsuccessful, and IBM instead made an agreement with Microsoft. Kildall died in tragic circumstances in 1994 at the young age of 52. He received a posthumous award from the Software Publishers Association in 1995 for his contri-butions to the microcomputer industry.

42.1   Licensing CP/M to IBM

Kildall lost out on the opportunity of a lifetime to supply the operating for the IBM personal computer to IBM. Don Estridge led the IBM team that developed the IBM personal computer, and the project was subject to an aggressive delivery schedule. Traditionally, IBM developed a full proprietary solution for its products, but due to time pressures, it decided to outsource the development of the microprocessor and the operating system.

The IBM team initially asked Bill Gates and Microsoft in Seattle to supply them with an operating system. Microsoft had signed a contract with IBM to supply a BASIC interpreter for the IBM PC, but they lacked the expertise in operating system development. Gates referred IBM to Gary Kildall at DRI, and the IBM team approached Digital Research with a view to licensing their CP/M operating system.

Digital Research was working on CP/M-86 for the Intel 16-bit 8086 microprocessor (introduced by Intel in 1978). IBM decided to use a lower cost and slower version of the 8086 (the Intel 8088), which was introduced in 1979, for its new personal computer.

IBM and Digital Research failed to reach an agreement on the licensing of CP/M for the IBM PC. The precise reasons for failure are unclear, but some immediate problems arose with respect to the signing of an IBM non-disclosure agreement during the visit. It is unclear whether Kildall actually met with IBM and whether there was an informal handshake agreement between both parties. However, there was no legal written agreement between IBM and DRI.

There may also have been difficulties with respect to the royalty fee being demanded by Digital Research, as well as difficulties in meeting the aggressive IBM delivery schedule (due to Digital Research's existing commitments to Intel).

Gates had been negotiating a Microsoft BASIC license agreement with IBM, and he saw a business opportunity for Microsoft. He offered to provide an operating system (later called PC/DOS) and BASIC to IBM on favorable terms. IBM accepted the offer, and the rest, as they say, is history. Gates was aware of the work done by Tim Patterson on a simple quick and dirty version of CP/M (called QDOS) for the 8086 microprocessor for Seattle Computer Products (SCP). Gates licensed QDOS for $50,000 and hired Patterson to modify it to run on the IBM PC (for the 8088 microprocessor).

Gates then licensed the operating system to IBM for a low per copy royalty fee. IBM called the new operating system PC/DOS, and Gates retained the rights to MS/DOS, which were used on IBM-compatible computers produced by other hardware manufacturers. MS/DOS later became the dominant operating system for the PC (eclipsing PC/DOS due to the open-architecture IBM PC and the growth of clones). Microsoft became a major corporation.

DRI released CP/M-86 shortly after IBM released PC/DOS. Kildall examined PC/DOS, and it was clear to him that it had been derived from CP/M. He was furious and met separately with IBM and Microsoft, but nothing was resolved. Digital

Research considered suing Microsoft for copying all of the CP/M system calls in DOS 1.0, as it was evident that Patterson's QDOS was a copy of CP/M. He considered his legal options, but it was not clear at that time what constituted infringement of copyright. There was no guarantee of success in any legal action against IBM, and it was potentially very expensive. Kildall threatened IBM with legal action, and IBM agreed to offer both CP/M-86 and PC/DOS. However, as CP/M was priced at $240 and DOS at $60, few PC owners were willing to pay the extra cost. CP/M was to fade into obscurity.

Perhaps, if Kildall had played his hand differently, Digital Research could have been the Microsoft of the PC industry. DRI was slow in developing the 16-bit operating system, which gave Patterson the opportunity to create his own version. IBM was under time pressure with the development of the IBM PC, and Kildall was unable to meet the IBM deadline. This resulted in IBM dealing with Gates instead of DRI. Further, the royalty fee demanded by Kildall for CP/M resulted in very low sales for his product, whereas if a more realistic price had been charged, then DRI may have made some reasonable revenue. Nevertheless, Kildall must have viewed Microsoft's actions as the theft of his intellectual ideas and technical inventions.

The IBM PC was introduced in 1981, and the first version of the operating system was compatible with Digital Research's CP/M operating system (as it essentially was CP/M). It managed floppy disks and files, input and output, and memory, and it contained an external command processor that interpreted user commands and allowed the user to interact with the system.

MS/DOS version 2.0 was introduced in 1983, and it was designed to support the 10 MB hard disk on the IBM PC/XT as well as provide support for device drivers. Microsoft had previously licensed XENIX (their commercial version of UNIX) from AT&T, and MS/DOS 2.0 was a move toward XENIX. It employed a hierarchical file system, and a unique path name identified each file (similar to XENIX). It provided limited multitasking for background print spooling. The hard disk on the XT helped to establish the IBM PC in the business marketplace.

The open architecture of the IBM PC led to the development of cheaper IBM compatible personal computers (clones of the IBM PC but cheaper). These rapidly gained market share, as it was difficult for IBM to compete on price. This resulted in huge demand for MS/DOS (which was the operating system for IBM compatibles and clones).

MS/DOS 3.0 was released in 1984, and it provided support for the IBM PC/AT, which had a 20 MB hard disk. Several versions of MS/DOS followed through the 1980s and 1990s and were used with Microsoft Windows 95 and Windows Millennium. Today, Microsoft Windows is the operating system used on personal computers, and MS/DOS is of historical interest.

Chapter 43
Office Software

Microsoft Office is a suite of office applications for the Microsoft Windows operating system. It consists of well-known programs such as Microsoft Word, which is a word processor; Microsoft Excel, which is a spreadsheet program; Microsoft PowerPoint, which is used to create slideshows for presentations; Microsoft Access which is a database management system for Windows; and Microsoft Outlook which is a personal information manager.

Microsoft's first Office application was a spreadsheet program initially called Multiplan when it was released in 1982. It was developed as a competitor to VisiCalc (Apple's spreadsheet program), and it renamed to *Excel* when it was released on the Macintosh in 1985. Excel is a spreadsheet program consisting of a grid of cells in rows and columns that may be used for data manipulation and arithmetic operations. It includes functionality for statistical, engineering, and financial applications, and it can display lines, histograms, and charts. It also provides support for user-defined macros.

Microsoft Word is the leading word processor, and the first version of the program was released on the MS/DOS operating system in 1983. It was designed for use with a mouse, and it provides *What you see is what you get* functionality. The first version of Word for Windows was released in 1989, and Microsoft Word began to dominate the market from the early 1990s.

Microsoft PowerPoint is a popular presentation program, and it enables the user to create a presentation consisting of several slides. Each slide may contain text, graphics, audio, movies, and so on. PowerPoint has made it easier to create presentations. It was originally developed for the Macintosh computer in 1987, and it was released for Windows in 1990.

The first version of *Microsoft Access* was released in 1992, and this database management system enables users to create tables, queries, forms, and reports. It includes a graphical user interface that allows users to build queries without knowledge of the query language. *Microsoft Outlook* is a personal information manager, and it is used mainly as an email application, but it also includes a calendar, task manager, note taking, and web browsing.

© Springer Nature Switzerland AG 2018
G. O'Regan, *The Innovation in Computing Companion*,
https://doi.org/10.1007/978-3-030-02619-6_43

The various Microsoft application programs such as Word, Excel, and PowerPoint were all available individually, until they were bundled together into the Microsoft Office suite in 1989.

43.1 Microsoft Excel

Microsoft Excel is a spreadsheet program, and it consists of a grid of cells in rows and columns that may be used for data manipulation and arithmetic operations. It includes functionality for statistical, engineering, and financial applications, and it has graphical functionality to display lines, histograms, and charts (Fig. 43.1).

This spreadsheet program was initially called Multiplan when it was released in 1982, and it was Microsoft's first Office application. It was developed as a competitor to Apple's VisiCalc, and it was initially released on computers running the CP/M operating system. It was renamed to *Excel* when it was released on the Macintosh in 1985, and the first version of Excel for the IBM PC was released in 1987.

It provides support for user-defined macros, and it also allows the user to employ Visual Basic for Applications (VBA) to perform numeric computation and report the results back to the Excel spreadsheet. Lotus 1-2-3 was the leading Spreadsheet tool of the 1980s, but Excel overtook it from the early 1990s.

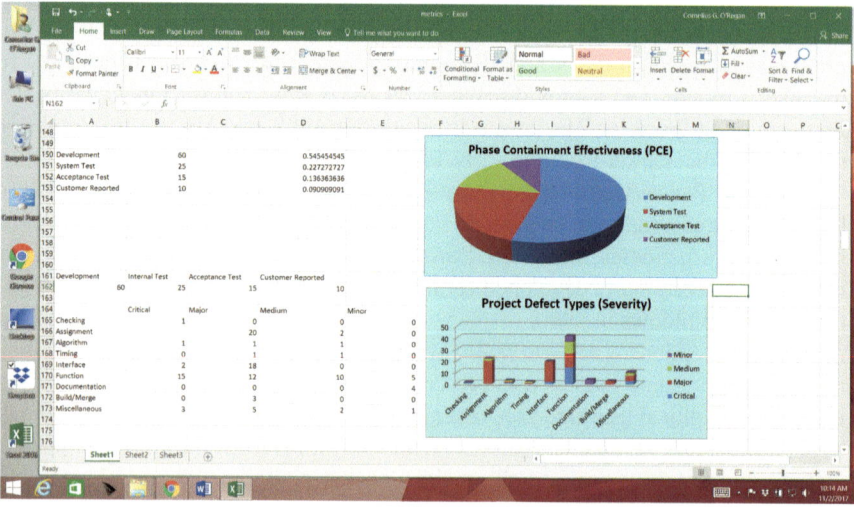

Fig. 43.1 Microsoft Excel screenshot. Used with permission from Microsoft

43.2 Microsoft PowerPoint

Microsoft PowerPoint is a popular presentation program that allows the user to create a presentation consisting of several slides. Each slide may contain text, graphics, audio, movies, and so on, and PowerPoint has made it easier to create and deliver presentations. The user may customize slideshows and show the slides in a different order from the original order. It has advanced features for animating text and graphics, video editing, and even broadcasting the presentation.

Microsoft PowerPoint was initially called Presenter, and Forethought Inc. originally developed it in 1987 for the Macintosh computer. Microsoft acquired Forethought for $14 million in 1987, and the first Windows version of PowerPoint was released in 1990. PowerPoint has many features to enable professional presentations to be made (Fig. 43.2).

43.3 Microsoft Word

Microsoft Word is used for word processing tasks such as creating and editing documents. Charles Simonyi and Richard Brodie developed it for the MS/DOS operating system in the early 1980s. Simonyi and Brodie were former Xerox PARC employees who had worked on the Xerox Bravo WYSIWYG GUI word processor (the first such word processor), and they joined Microsoft in 1981. The first version of Microsoft Word was released in 1983.

WordStar was the leading word processor at the time, and it took some time for Microsoft Word to gain popularity. Word was designed for use with a mouse, and it

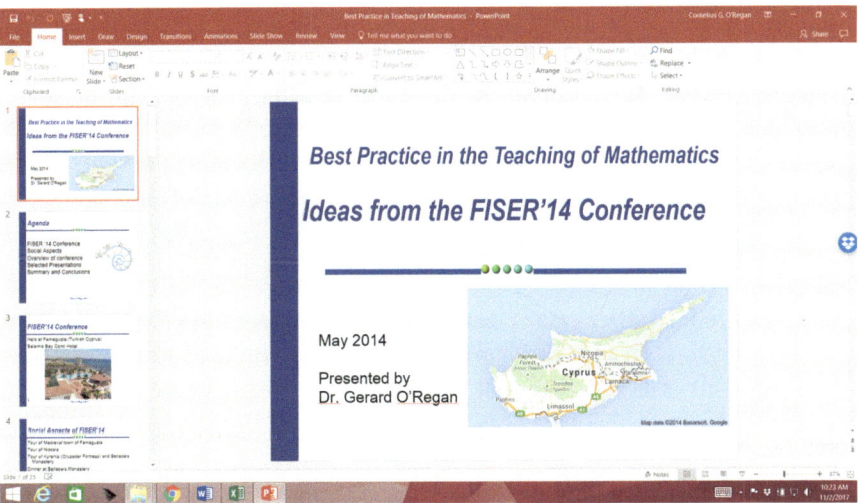

Fig. 43.2 Microsoft PowerPoint screenshot. Used with permission from Microsoft

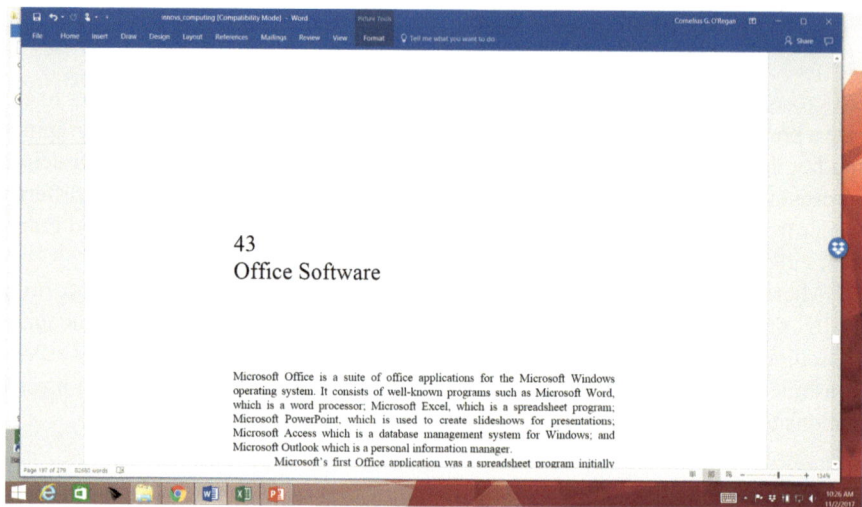

Fig. 43.3 Microsoft Word screenshot. Used with permission from Microsoft

provided *What you see is what you get* (WYSIWYG) functionality. Microsoft continued to improve the product, and it was ported to the MAC operating system in 1985. The first version for Windows was released in 1989, and Word began to dominate the word processing market shortly after the release of Windows 3.0. (Fig. 43.3)

43.4   Microsoft Access and Outlook

Microsoft Access is a database management system that allows users to create tables, queries, forms, and reports and connects them together with macros. It includes a graphical user interface that allows users to build queries without knowledge of the query language, or the user can create the query using the SQL database query language.

Microsoft Outlook is a powerful email program and a personal information manager. It allows user to schedule meetings and to book meeting rooms and other resources, and the main Outlook sections include Mail, Calendar, Contacts, Tasks, Notes, and Journal. Users may create and send email messages and manage their emails by creating email rules; they may create email auto-reply messages to automatically reply when they are out of the office; manage meetings, events, and appointments; maintain and manage contacts; and define tasks that the user needs to perform (including their priority).

Chapter 44
Open-Source Software

The two most common categories of software licenses that may be granted under copyright law are those for *proprietary software* and those for *free open-source software* (FOSS). The rights granted to the licensee are quite different for each of these categories, where the user has the right to copy, modify, and distribute (under the same license) software that has been supplied under an open-source license, whereas proprietary software typically does not grant these rights to the user.

The licensing of proprietary software typically gives the owner of a copy of the software the right to use it (including the rights to make copies for archival purposes). The software may be accompanied with an end-user license agreement (EULA) that may place further restrictions on the rights of the user. There may be restrictions on the ownership of the copies made and on the number of installations allowed under the term of the distribution. The ownership of the copy of the software often remains with the copyright owner, and the end user must accept the license agreement to be able to use the software.

Open-source software (OSS) is software that is freely available under an open-source license to study, change, and distribute to anyone for any purpose. Free and open-source licenses are often divided into two categories depending on the rights to be granted in distribution of the modified software. The first category aims to give users unlimited freedom to use, study, and modify the software and, if the user adheres to the terms of an open-source license such as GNU General Public License (GPL), the freedom to distribute the software and any changes made to it. The second category of open-source licenses gives the user permission to use, study, and modify the software but not the right to distribute it freely under an open-source license (it could be distributed as part of a proprietary software license).

Open-source software has become popular in recent times, and it allows software developed by others to be used (*under an open-source license*) in the development of applications. This modern approach to software development allows the source code to be published, and thousands of volunteer software developers from around the world participate in developing and improving the software code. The idea is that the source code is not proprietary and that it is freely available for software

© Springer Nature Switzerland AG 2018
G. O'Regan, *The Innovation in Computing Companion*,
https://doi.org/10.1007/978-3-030-02619-6_44

developers to use and modify as they wish. One useful benefit is that it may potentially speed up development time thereby shortening time to market.

The roots of open-source development are in the Free Software Foundation (FSF). This is a nonprofit organization founded by Richard Stallman (O'Regan 2013b) to promote free software, and it has developed a legal framework for open-source software development.

The Linux operating system is a well-known open-source product, and other products include MySQL, Firefox, and Apache HTTP server. Google introduced its open-source Android operating system in late 2007, which is the dominant operating system for smartphones and tablets. The quality of software produced by the open-source movement is good, and defects are generally identified and fixed faster than with proprietary software.

44.1   Free Software Foundation

Richard Stallman (Fig. 44.1) is the prophet of the free software movement, and he launched the Free Software Foundation (FSF) in 1985. Stallman became interested in computers at high school, after spending a summer working at IBM's Scientific Center in New York. He joined the Artificial Intelligence Laboratory at MIT as a programmer, and he later became a critic of restricted computer access at the lab. He believed that software users should have the freedom to share software with others and to be able to study and make changes to the software that they use. He left his position at MIT to launch the free software movement, and he explains his concept of free software as:

> "Free software is a matter of liberty, not price. To understand the concept, you should think of free as in free speech, not as in free beer."

He launched the GNU project in 1984, which is a free software movement and involves the participation of volunteer software programmers from around the

Fig. 44.1 Richard Stallman. Creative commons

world. He formed the Free Software Foundation (FSF) to promote the free software movement, and he is the non-salaried president of the organization. FSF has developed a legal framework for the free software movement, which provides a legal means to protect the modification and distribution rights of free software. The meaning of the term *free software* in defined in the GNU manifesto, and he lists four key freedoms essential to software development (Stallman 2002), and a program is termed *free* if it satisfies these properties. These are:

1. Freedom to run the program for any purpose.
2. Freedom to access, study, and improve the code and to modify it to suit your needs.
3. Freedom to make copies of the program and to redistribute them to others.
4. Freedom to distribute copies of the modified program so that others can benefit from your improvements.

The GNU project uses software that is free for users to copy, edit, and distribute. It is free in the sense that users can change the software to fit individual needs. Stallman has written many essays on software freedom and is a key campaigner for the free software movement. The legal framework for the free software movement provides protection to the modification and distribution rights of free software. Stallman introduced the concept of *copyleft*, which is a form of *licensing of free software*. It makes a program or product free and requires that all modified or extended versions of the program are also free.

Copyright law grants exclusive rights to the copyright holder for a period of time to reproduce the work, to extend the work, to distribute copies of the work, and to perform or display the work. The duration of the copyright may be more than 50 years after the author's death. Any unauthorized use of the author's work is termed a *copyright infringement*, and the copyright owner may take legal action to deal with an infringement.

A *patent is a form of intellectual property that grants exclusive rights to the inventor* to the commercial benefits of the invention for a limited time duration. The invention must be novel and more than just an obvious next step from existing inventions. The rights are granted to the inventor in a particular country, and the time period is typically 20 years. The rights granted under patent law are exclusive and prevent others from the unauthorized use of the invention. Patents may be legally enforced by civil lawsuits, and *patent licensing agreements* are legal contracts where the patent owner grants rights to others to use the invention.

Stallman has argued against intellectual property such as patent law and copyright law. He has argued against patenting software ideas, stating that a patent is an absolute monopoly on the use of an idea. He states that while 20 years may not seem like a long period of time, that in the software field, it is essentially a generation, due to the pace at which technology changes in the world. Further, *patents act a barrier to competition and lead to monopolies*. They make it difficult for new companies to enter a marketplace, due to the restrictions and costs associated with the licensing of patents. In recent times, we have seen large companies acquire others for their intellectual property (e.g., the Google acquisition of Motorola Mobility was due to the latter's valuable collection of patents), and today there are major intellectual property wars in the corporate world.

Stallman argues that copyright law places Draconian restrictions on the public and takes away freedoms that they would otherwise have. They protect the businesses of the copyright owner, and he suggests that alternative approaches should be considered in the digital age.

44.2   GNU

GNU has a recursive definition and it stands for GNU's Not UNIX. The founding goal of the GNU project was to develop *a sufficient body of free software to get along without any software that is not free*. Stallman announced his plan to create a free operating system called GNU in 1983, and he published the GNU manifesto in 1985. The manifesto aimed to gain support from other developers for the project, and the initial goal was the development of a new operating system that would be compatible with UNIX.

He created several tools for GNU, including the EMACS text editor, the GCC compiler, the gdb debugger, and the gmake build automator. A Finnish student, Linus Torvalds, used the GNU tools to create the Linux kernel in 1992, and the resulting free operating system platform is known as GNU Linux or just Linux.

Chapter 45
Object-Oriented Paradigm

Object-oriented design (OOD) is a design method that models the system as a set of cooperating objects rather than as a set of functions and where the individual objects are viewed as instances of a class. Object-oriented design is concerned with the object-oriented decomposition of the system, and it involves defining the required objects and their interactions to solve the problem. The system state is decentralized with each object managing its own state information. The objects have a collection of attributes that define their state and operations that act on the state. The data in the object is hidden, and the only access to the data is with the operations.

The difference between a class and an object may be seen from the example that walls and windows are classes, whereas individual doors and windows are objects. A class is a set of objects (rather than an individual object), and all members of the class share the same attributes, operations, and relationships. A class may represent a software thing or a hardware thing.

A class may inherit its behavior from one or more super-classes, with the class definition setting out the differences between the class and its super-classes. The communication between objects is done by exchanging messages (in practice, an object calls a procedure associated with another object).

An object is a **"black box"** that sends and receives *messages*. A black box consists of *code* (computer instructions) and *data* (information which these instructions operate on). The traditional way of programming kept code and data separate. For example, functions and data structures in the C programming language are not connected. However, in the object-oriented world, code and data are merged into a single indivisible thing called an *object*.

The reason that an object is called a black box is that the user of an object never needs to look inside the box, since all communication to it is done via messages. Messages define the *interface* to the object. Everything that an object can do is represented by its message interface. Therefore, there is no need to know anything about what is inside the black box (or object) to use it. The access to an object is only through its messages while keeping the internal details of the object private.

© Springer Nature Switzerland AG 2018
G. O'Regan, *The Innovation in Computing Companion*,
https://doi.org/10.1007/978-3-030-02619-6_45

Table 45.1 Object-oriented paradigm

Feature	Description
Class	A class defines the abstract characteristics of a thing, including its attributes (or properties) and its behaviors (or methods). The members of a class are termed objects
Object	An object is an instance of a class with its own set of attributes. The set of values of the attributes of an object is called its state
Method	The methods associated with a class represent the behaviors of the objects in the class
Message passing	Message passing is the process by which an object sends data to another object or asks the other object to invoke a method
Inheritance	A class may have subclasses (or children classes) that are more specialized versions of the class. A subclass inherits its attributes and methods of the parent class. The programmer may create new classes from existing classes, and the derived classes inherit the methods and data structures of the parent class
Encapsulation (information hiding)	The internals of an object are kept private, and may not be accessed outside of the object. The details of how a class works are hidden, and a clearly specified interface around its services is provided
Abstraction	Abstraction simplifies complexity by modeling classes and removing all unnecessary detail
Polymorphism	Polymorphism is behavior that varies depending on the class in which the behavior is invoked. Two or more classes may react differently to the same message

It is important to understand the relationship between the software to be designed and its external environment. This may be done with the unified modeling language (UML) to develop models that show the other systems in its environment and the interaction between them. This leads to the architectural design where the major components of the system and their interactions are defined. The UML diagrams help in identifying the objects and operations in the system and the relationships between them. The main features of the object-oriented paradigm are described in Table 45.1.

The traditional view of programming is *procedural*, where a program is viewed as a collection of functions (i.e., a list of instructions to be performed on the computer). Object-oriented programming is a paradigm shift in programming, where a program is viewed as a collection of *objects* that act on each other. Each object is a black box that is capable of sending and receiving messages and processing data, and it may be viewed as an independent entity/actor with a distinct role or responsibility.

The traditional way of programming kept code and data separate. The functions and data structures in the C programming language are separate, whereas in the object-oriented world of C++, code and data are merged into a single indivisible thing called a *class* (and *objects* of the class).

The user of an object never needs to look inside the black box, and messages define the *interface* to the object. This approach is termed *information hiding*[1] and was developed by Parnas in the early 1970s (Parnas 1972).

The origins of object-oriented programming go back to the invention of Simula 67 at the Norwegian Computing Research Center (NR)[2] in the late 1960s. Simula 67 introduced the notion of a class and instances of a class[3]. The Smalltalk language was developed at Xerox PARC in the mid-1970s, and it introduced the term *object-oriented programming* for the use of objects and messages as the basis for computation.

Object-oriented programming became the dominant paradigm in programming from the late 1980s. Its proponents argue that it is easier to learn and simpler to develop and maintain. Its growth in popularity was helped by the rise in popularity of Graphical User Interfaces (GUI), as the development of GUIs is especially suited to object-oriented programming.

The Java and C++ programming languages have become very popular, and object-oriented features have been added to many existing languages including COBOL and FORTRAN. There are many available textbooks on object-oriented design and development (e.g., (Weisfield 2013)).

---

[1] Information hiding is a key contribution by Parnas to computer science. He has also done work on mathematical approaches to software quality using tabular expressions (O'Regan 2017b).

[2] The inventors of Simula 67 were Ole-Johan Dahl and Kristen Nygaard.

[3] Dahl and Nygaard were working on ship simulations and were attempting to address the huge number of combinations of different attributes from different types of ships. Their insight was to group the different types of ships into different classes of objects, with each class of objects being responsible for defining its own data and behavior.

Chapter 46
Personal and Home Computers

The introduction of the IBM personal computer was a paradigm shift in computing, in that it placed computing power directly in the hands of millions of people. The previous paradigm was that an individual user had limited control over a large mainframe computer, with the system administrators controlling the access privileges of the individual users.

The invention of the microprocessor was a revolution in computing (see Chap. 38), and it led inexorably to the development of home and personal computers. The early home computers included the MITS Altair 8800, the Apple I and II computers, the Commodore PET and 64 computers, the Atari 400 and 800 computers, the Sinclair ZX81 and ZX Spectrum, and the Apple Macintosh.

IBM introduced its IBM Personal Computer (or PC) in 1981 as a machine to be used by small businesses and users in the home (Fig. 46.1). Its strategy at the time was to get into the home computer market as quickly as possible, in order to break the dominance of Commodore, Atari, and Apple. The IBM PC was priced at $1565, and it was affordable to computer users. It offered 16 kilobytes of memory (expandable to 256 kilobytes), a floppy disk, and a monitor. It was an immediate success and became the industry standard.

IBM assembled a small team of 12 people led by Don Estridge, and the chief designer was Lewis Eggebrecht. Their goal was to design and develop the IBM PC within 1 year, and as time to market was a key driver, they built the machine with *off-the-shelf* parts from several equipment manufacturers. They had intended using the IBM 801 processor developed at the IBM Research Center in Yorktown Heights but decided instead to use the Intel 8088 microprocessor, which was inferior to the IBM 801. They chose to outsource the development of the PC/DOS operating system (see Chap. 41) from Microsoft rather than developing their own operating system.

The unique IBM elements in the personal computer were limited to the system unit and keyboard. The team decided on an open architecture so that other manufacturers could produce and sell peripheral components and software without purchasing a license. They published the IBM PC Technical Reference Manual, which

© Springer Nature Switzerland AG 2018
G. O'Regan, *The Innovation in Computing Companion*,
https://doi.org/10.1007/978-3-030-02619-6_46

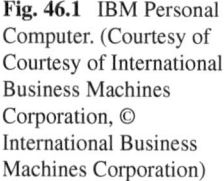

Fig. 46.1 IBM Personal
Computer. (Courtesy of
Courtesy of International
Business Machines
Corporation, ©
International Business
Machines Corporation)

included the complete circuit schematics, the IBM ROM BIOS source code, and
other engineering and programming information.

Don Estridge (Fig. 46.2) led the project, and they developed a prototype machine
in 4 months, and the completed personal computer was available within the 1-year
time frame. The IBM personal computer overtook Apple as the best-selling personal
computer in 1983.

The open architecture led to a new industry of *IBM-compatible* computers, which
had all essential features of the IBM PC but were cheaper. The terms of the licensing
of PC/DOS operating system gave Microsoft the rights to license its MS/DOS operating system used on the IBM compatibles, and this led inexorably to the rise of the
Microsoft Corporation. The IBM Personal Computer XT was introduced in 1983.
This model had more memory, a dual-sided diskette drive, and a high-performance
fixed-disk drive. It was followed by the introduction of the Personal Computer/AT
in 1984.

The development of the IBM PC meant that computers were now affordable to
ordinary users, and this led to a huge consumer market for personal computers and
software. It led to the development of business software such as spreadsheets and
accountancy packages, banking packages, programmer developer tools such as
compilers and specialized editors, and computer games. The introduction of the
personal computer represented a paradigm shift in computing, and it placed computing power directly in the hands of millions of people.

IBM had traditionally produced all components for its machines. However, due
to the aggressive time schedules associated with the IBM PC, it decided to outsource

Fig. 46.2 Don Estridge. (Courtesy of Courtesy of International Business Machines Corporation, © International Business Machines Corporation)

the production of components to other companies. This proved to be a *major error* as it outsourced the production of the processor chip to a company called Intel[1], and the development of the PC Disk Operating System (PC/DOS) was outsourced to a small company called Microsoft[2]. These companies later become technology giants.

46.1   Home Computers

The Xerox Alto computer was developed at Xerox PARC, and it pioneered several key concepts in personal computing such as the use of a graphical user interface and the mouse. It was essentially a small minicomputer rather than a personal computer, and it was unlike modern personal computers in that it was not based on the microprocessor.

The first real home computer was the MITS Altair 8800, and the prototype machine was available in October 1974 (Fig. 46.3). The cover page of the January 1975 edition of Popular Electronics featured an early design of the Altair 8800, and this publicity helped in generating sales that vastly exceeded expectations. It included a home computer kit version (which was assembled by the customer) and a fully assembled version.

---

[1] Intel was founded by Bob Noyce and Gordon Moore in 1968.

[2] Microsoft was founded by Bill Gates and Paul Allen in 1975.

Fig. 46.3 MITS Altair
home computer

The home kit included assembly instructions, a metal case, a front panel with switches, a power supply, a motherboard with expansion slots, and various cards to plug into the expansion slots, as well as any other components required to build the computer. The actual assembly was quite a challenge as it involved careful soldering and assembly. There was no actual keyboard or monitor, which meant that the task of programming the machine was nontrivial and required the user to program in machine language and watch the LEDs on the front panel to get the results. Several expansion cards (e.g., for keyboard, monitor, and data storage) were soon released, and this made it easier to use the machine. The Altair 8800 used the 8-bit Intel 8080 microprocessor, which was introduced in 1974.

Bill Gates and Paul Allen developed a BASIC interpreter for the Altair 8800, and the 4 k/8 k versions of BASIC were released in July 1975; Gates and Allen formed Microsoft later that year.

Many home computers have been developed since then including the Apple I and Apple II computers, which were released in 1977. The GUI-driven Apple Macintosh was released in 1984 (see Chap. 5). The Commodore PET and Commodore 64 were released in 1977 and 1982, respectively (see Chap. 18). The Atari 400 and 800 were introduced in 1979. The Sinclair ZX81 and Sinclair ZX Spectrum were released in 1981 and 1982, respectively.

Chapter 47
Robotics

The first use of the term "robot" was by the Czech playwright, Karel Capek, in his play *Rossum's Universal Robots*, which was first performed in Prague in 1921. The word *robot* is a Czech word for forced labor, and the play explores whether it is ethical to exploit artificial workers in a factory and how the robots should respond to their exploitation. Capek's robots looked and acted like humans and were created by a chemical process. They were not mechanical or metal in nature, and Capek rejected the idea that machines created from metal could think or feel.

Asimov wrote several stories about robots in the 1940s including the story of a robotherapist. He predicted the rise of a major robot industry, and he also introduced a set of rules (or laws) for good robot behavior. These are known as the three *Laws of Robotics* (Table 47.1), and Asimov later added a fourth law.

Robots have been very effective at doing clearly defined repetitive tasks, and there are many sophisticated robots in the workplace today. These are industrial manipulators that are essentially computer-controlled "arms and hands." They can improve the quality of life for workers as they can free human workers from performing dangerous or repetitive jobs. They provide consistently high-quality products and can work tirelessly 24 hours a day. This helps to reduce the costs of manufactured goods thereby benefiting consumers. The term *robot* is defined by the Robot Institute of America as:

> A *re-programmable, multifunctional manipulator designed to move material, parts, tools, or specialized devices through various programmed motions for the performance of a variety of tasks.*

George Devol was awarded the patent for the first industrial robot (*Unimate*), and he played an important role in the foundation of the modern robotics industry. He is regarded (with Joseph Engelberger) as one of the fathers of robotics.

He applied for a patent on Programmed Article Transfer in 1954, and the goal of this invention was to perform repeated tasks with greater precision and productivity than a human worker. The patent was concerned with automatic operation of machinery, including handling and an automated control apparatus. It introduced

© Springer Nature Switzerland AG 2018
G. O'Regan, *The Innovation in Computing Companion*,
https://doi.org/10.1007/978-3-030-02619-6_47

Table 47.1 Asimov's Laws of Robotics

Law	Description
Law zero (humanity)	A robot may not injure humanity or, through inaction, allow humanity to come to harm
Law one (human safety)	A robot may not injure a human being or, through inaction, allow a human being to come to harm, unless this would violate a higher-order law
Law two (slave)	A robot must obey orders given by human beings, except where such orders would conflict with a higher-order law
Law three (survival)	A robot must protect its own existence as long as such protection does not conflict with a higher-order law

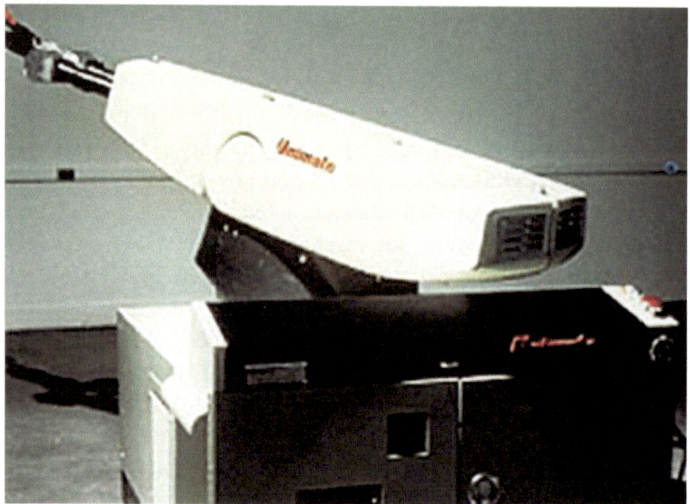

Fig. 47.1 Unimate in 1960s

the concept of *universal automation*, and the term *Unimate* was coined. This was the first patent for a digitally operated programmable robot arm, and it led to the foundation of the robotics industry (Fig. 47.1).

Engelberger and Devol founded the first robotics company, *Unimation Inc.*, in 1956. This was the largest robotics company in the world for many years. They initially developed a material-handling robot and robots for welding, and they installed the first industrial robot (Unimate) on a General Motors (GM) assembly line in 1961. This robot was used to lift hot pieces of metal from a die-casting machine and to stack them. The story of Unimation Inc. is told in (Munson 2011).

These robots were very successful and reliable and saved General Motors money by replacing staff with machines. Other automobile companies followed GM in purchasing Unimate robots, and the robot industry continues to play a major role in the automobile sector.

A Unimate robot appeared on *The Tonight Show* hosted by Johnny Carson in 1966, and the robot poured a beer and sank a golf putt. Unimate was named by *Popular Mechanics* magazine in 2005 as one of the top 50 inventions of the past 50 years, and Devol has received several awards for his contributions to robotics.

An intelligent robot is a machine that can extract information from its environment and use knowledge about the physical world to move safely in a meaningful and purposeful manner. The robot senses its environment and acts in a rational manner to achieve the defined goals. Robots require good engineering and science to be effective, and the engineering features include:

– Sensors.
– Effectors/actuators.
– Locomotion System.
– On-board computer system.
– Controllers.

The sensors provide limited and crude information to the robot on its environment, where the environment may contain lots of potentially useful information. These may include visual sensors such as cameras, auditory sensors that detect atmospheric vibrations and extract information from the sounds, and olfactory sensors (electronic nose) to detect odors or flavors and sensors that aim to develop a sense of touch in robots (Fig. 47.2).

The robot exists in its sensor space (i.e., the set of all possible values of its sensor readings), and robot sensors are different from biological sensors. The robot's interpretation of the world is based on its sensory perceptions, and its response to the sensory data may involve the use of its effectors (e.g., leg, wheel, arm) to act and interact with the environment.

The area of computer vision has made significant progress in recent times with the Stanford Cart developed in the mid-1970s and the Carnegie Mellon Rover and its successors from the early 1980s. The Stanford Cart was a simple buggy with a video camera, and Han's Moravev's version in the late 1970s could navigate slowly around a room with obstacles (without human intervention) in a controlled environment. The Carnegie Mellon Rover and its successors have been very successful,

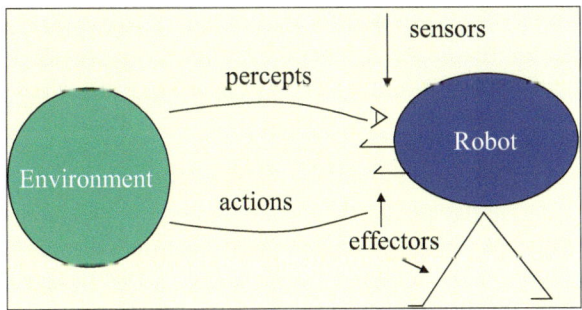

Fig. 47.2 Robot interacting with environment

Table 47.2 Advantages of robots

No	Advantage
1.	Robots can do repetitive work 24× 7 × 365
2.	Robots can do tasks that are too dangerous for humans
3.	Robots can operate machinery to a much higher level of precision than humans
4.	A robot may be able to perform tasks that are impossible for humans

Table 47.3 Disadvantages of robots

No	Disadvantage
1.	Robots can make incorrect decisions in emergencies
2.	Robots have limited movement and vision
3.	Robots are costly and require programming and training

with the CMU Navlab 5 vehicle traveling from Pittsburg to Los Angeles in 1995 with an autonomous driving percentage of 98.2%.

Robots have been applied to many areas including assembly and manufacturing, industrial manipulation and material handling, hazardous environments, space and underwater exploration in unmanned vehicles, education, remote environments, medical science, virtual reality, and the entertainment industry. The advantages of robots include (Table 47.2).

The disadvantages of robots include (Table 47.3).

47.1   Robots and Ethics

As more and more sophisticated machines and robots are created, it is, of course, essential that intelligent machines behave ethically and have a moral compass to distinguish right from wrong. It remains an open question as to how to teach a robot right from wrong, and in view of recent progress that has been made in the AI field, the time is approaching where machines will need to routinely make ethical decisions.

For example, it is reasonable to expect that driverless cars (self-driving vehicles) will be common on the road in the next 10–20 years. A driverless car is a vehicle that can sense its environment and navigate a route without human intervention. Suppose a self-driving vehicle is traveling on a road and two children roll off a grassy bank on to the road and there is no time for the vehicle to brake. However, if the vehicle swerves to the left, it can avoid the children but hit an oncoming motorbike. *Which decision should the car make, and how should it make such a decision?*

This is a variant of the *trolley problem*, which is a famous thought experiment in ethics. A train is rushing down a track out of control as its brakes have failed.

Disaster lies ahead as five people are tied to the track and will perish in the absence of action. There is sufficient time to flick the points and divert the train down a side track where there is one man tied to the track. Is it ethical to divert the train to do this? Most people would be inclined to take the view that it is the best (least worst) possible outcome.

There is a controversial variant of the problem where the train is rushing toward five people, and you are standing on top of a footbridge overlooking the track next to a man with a very bulky rucksack. The only way to save the five people is to push the man to his doom, as his rucksack will block the train and save the five. Is it ethical to deliberately kill or sacrifice another human being to save five others? Most people would say no to this deliberate killing, but it would be valid in the utilitarian school of ethics, which seeks to maximize happiness in the world.

Even though the trolley problem is a thought experiment, it is conceivable that a driverless car will face situations where a moral choice must be made (e.g., who to harm or injure such as pedestrians, passengers, or driver). Clearly, this raises the importance of the type of ethics that are programmed into the car and who is to decide what ethics are programmed into a car.

Teaching ethics may involve programming in certain principles, and then the machine learns from scenarios on how to apply the principles to new situations. There is a need for care with machine learning as the machine may learn the wrong lessons, or as its learning evolves, it may not be possible to predict its behavior in the future.

Further questions arise as to who is to be held accountable in the event of a machine making incorrect or unethical decisions. For further information on the feasibility of teaching ethics to robots, see the fascinating BBC article "Can we teach robots ethics?" (BBC Magazine 2017).

47.2 Robots and Intelligence

There has been a gradual improvement in the capability of robots due to the exponential advances in computer technology (due mainly to Moore's Law which was discussed in Chap. 32). However, despite the amazing advances in speed and processing power of computers, there has been no significant advance in the intelligence of machines. Machines and robotic intelligence is currently very low on the evolutionary scale, and robots lack basic common sense.

There have been some impressive results in artificial intelligence such as the Deep Blue chess program defeating Gary Kasparov, the world champion in chess, which have given the impression that machine intelligence will soon be within reach. However, whenever such results are analyzed, it is clear that they work only in very limited situations and that the machine cannot do anything else apart from what has been programmed to in the limited domain. For example, Deep Blue can run a simulation of reality for playing chess, but it is incapable of running simulations of any other reality.

The human brain is completely different from a digital computer, and it consists of a vast neural network of 100 billion neurons, with each neuron connected to about 10,000 others. Neural networks use a bottom-up approach to learning by experience rather than directly encoding the rules of intelligence. Neural networks reinforce neural pathways whenever a correct decision is made, which is done by changing the strength of the electrical connection between the neurons every time that it performs the task correctly.

Neural networks operate on quite different principles from digital computers, in that the human brain is massively parallel with 100 billion neurons operating at the same time and with each neuron performing some tiny act of computation. Further, learning in a neural network involves rewiring the brain and strengthening the links between neurons. The brain is not digital since a neuron may be analog as well as digital, whereas the transistors used in digital computers are either 1 or 0.

There has been limited progress made on machine (silicon) consciousness, where consciousness is interpreted as awareness of self and the environment, as well as the ability to make plans for the future (including running simulations of the future). Robots currently have a very limited understanding of their environment and lack the ability to fully recognize objects holistically in the environment. They have sensors to sense the environment, but they currently lack the pattern recognition capabilities of humans that allow a human to immediately recognize and understand objects in their environment.

There are questions as to whether machine intelligence is desirable and whether intelligent robots may be trusted to serve mankind. That is, the question is how benevolent and sociable intelligent robots will be if they are created in the future and whether they will wish to help mankind or will they wish instead to destroy the human race. However, it may be possible to design robots that will evolve over time into benevolent and sociable entities that will serve the human race.

It must be recognized that it has taken millions of years for humans to evolve to their current level of intelligence and that it will take significant time (probably several hundred years or longer) for machine intelligence to evolve to reach human levels.

Chapter 48
Smartphones and Social Media

Smartphones arose as the outcome of the marriage of the existing mobile phone technology and PDA technology, and they contain advanced computing capabilities that are attractive to users. Today, the smartphone is ubiquitous and is widely used around the world.

The introduction of the smartphone facilitated a major growth of social networking, as users could communicate news events or update their personal information in real time. Social networking sites such as Facebook and Twitter have transformed human communication.

Social media involves the use of computer technology that allows the creation and exchange of user-generated content. These web-based technologies allow users to collaborate to discuss and modify user created content. It has led to major changes in communication between individuals, communities, and organizations.

Facebook helps users to keep in touch with friends and family, and it allows them to share their opinions on what is happening around the world. Users may upload photos and videos, express opinions, exchange messages, and keep their community of friends up to date on important events in their lives. It has become an important channel for young people to discuss their aspirations for the future, as well as their grievances with society and the state. It has become an effective tool for protest and social revolution.

Twitter has become an effective way to communicate the latest news, and its effectiveness as a communication tool increases as the number of a person's followers grows. It allows a person or an organization to determine what people are saying about it, including their positive or negative experiences. This allows organizations and individuals to directly interact with their followers, and it is a powerful way to engage the audience and to make people feel heard.

© Springer Nature Switzerland AG 2018
G. O'Regan, *The Innovation in Computing Companion*,
https://doi.org/10.1007/978-3-030-02619-6_48

48.1 The Smartphone

A smartphone is more than a mobile device for making and receiving calls, and it is essentially a touch-based computer on a phone, which comes with its own keyboard, operating system, Internet access, and third-party applications. It provides many other features such as a camera, maps, calendar, alarm clock, and games.

IBM (in a joint venture with BellSouth) introduced an early precursor of today's smartphones in 1993. This was the IBM Simon, and it included voice and data services. It acted as a mobile phone, a PDA, and a fax machine, and it included a touch screen that could be used to dial numbers. It could send faxes and emails as well as making or receiving call. It included applications such as an address book, calendar, and calculator.

John Sculley coined the term *personal digital assistant* (PDA), and Apple introduced the first PDA, the Newton, in 1993. The Newton included some nice features including limited handwriting recognition abilities (the early versions of the product misread characters). Xerox PARC had created a prototype PDA, the Dynabook, in the 1970s, but it was not commercialized.

A PDA allows a large amount of data to be stored on a small handheld device. Palm introduced an early PDA device, the Palm Pilot 1000, in 1996, and it influenced the use of mobile data by business users. The Palm Pilot started the PDA industry, and it included 128Kb of memory and 16 MHz of processing power. It had better handwriting recognition capabilities than the Newton and a graphical user interface (GUI).

The Nokia 9000 Communicator was released in 1996, and it combined the features of a PDA and a mobile phone. It included a physical QWERTY keyboard, and it included features such as email, calendar, address book, and calculator. However, it did not provide the ability to browse the web, and a color display was introduced in the Nokia 9120 in 1998.

Qualcomm introduced its pdQ smartphone in 1999, and this phone combined a Palm PDA with Internet connectivity capabilities. Research In Motion (RIM) released its first Blackberry devices in 1999, and these provided secure email communication into a single inbox. Samsung's first smartphone was the Samsung SPH-I300, which was released in 2001, and this Palm-powered smartphone is a distant ancestor of today's smartphones. Samsung introduced its SGH i607 smartphone in 2006, and this Window's powered phone was inspired by Research in Motion's Blackberry phone.

Smartphone technology continued to evolve through the early 2000s, and Apple introduced its revolutionary *i*Phone in 2007. This Internet-based multimedia smartphone included a touch screen and features such as a video camera, email, web browsing, text messaging, and voice. The *i*Phone had a 3.5 inch 480 × 320 touch screen, a QWERTY keyboard, and 4GB of storage. Apple developed its own operating system, *i*OS, for the *i*Phone.

Google introduced its open-source Android operating system in late 2007, and the first Android phone was introduced in late 2008. Android is the dominant

Fig. 48.1 Apple iPhone 4

operating system for smartphones and tablets, with *i*OS used on Apple's products. The Samsung Instinct was released in 2008, but it was based on an operating system developed by Samsung. Although its touch screen operating system was not in the same league as Apple's *i*OS, it became a competitor to Apple's *i*Phone.

Apple's *i*Phone 4 (Fig. 48.1) was introduced in 2010, and this powerful smartphone has a 3.5 inch 960 × 640 screen and a 5 megapixel camera. The touch screen-enabled Samsung Galaxy S Android smartphone was launched in 2010, and the Samsung Galaxy S series of smartphones have been very successful and have become a major competitor to Apple's *i*Phone.

48.2 The Facebook Revolution

Facebook is the leading social networking site (SNS), and its mission is to make the world more open and connected. It enables users to keep in touch with friends and family and to share their opinions on what is happening around the world. Users may upload photos and videos, express opinions and ideas, and exchange messages. Facebook is very popular with advertisers as it allows them to easily reach a large target audience.

Mark Zuckerberg (Fig. 48.2) founded the company in 2004 while he was a student studying psychology at Harvard University. Zuckerberg was interested in programming, and he had already developed several social networking web sites for his fellow students including *Facemash* which could be used to rate the attractiveness of a person and *Coursematch* which allowed students to view people enrolled in a particular course.

Fig. 48.2 Mark
Zuckerberg

Zuckerberg launched *The Facebook* (thefacebook.com) at Harvard in February 2004, and over a thousand Harvard students had registered on the site within the first 24 h. Over half of the Harvard student population had a profile on Facebook within the first month. The membership of the site was initially restricted to students at Harvard, then to students at the other universities in Boston, and then to students at the other universities in the United States. Its membership was extended to international universities from 2005.

The use of Facebook was extended to anyone with an email address from 2006, and the number of registered users began to increase exponentially. The number of registered users reached 100 million in 2008 and reached 2 billion in 2017. It is now one of the most popular web sites in the world.

Facebook's business model is distinct from that of a traditional business in that it does not manufacture or sell any products. Its revenue is mainly from advertisements, and advertisements targeted to its over 2 billion users based on their specific interests. Facebook is essentially selling its users to advertisers (i.e., the users are the product), and the users do all the work. Facebook collects data about them (e.g., age, gender, location, education, work history, and interests) and classifies and categorize them, so that it can target advertisements that will potentially be of interest to them. This means that the advertisements are targeted to the right audience.

Social media have become important communication channels for educated young people to discuss their aspirations for the future, as well as their grievances with society and the state. The effectiveness of Facebook as a tool for protests and revolution is evident in the relatively short protests that culminated in the resignation of President Hosni Mubarak of Egypt in 2011.

Egypt has a young population with roughly 60% of the population under the age of 30, and the country has faced many challenges since independence such as improving education and literacy for its young population, as well as finding jobs for its citizens.

Facebook provided a platform for Egyptian youth to discuss issues such as unemployment, low wages, police brutality, and corruption. Young Egyptians set up groups on Facebook to discuss specific issues (e.g., a group that aimed to provide solidarity with striking workers was set up). Further momentum for revolution followed the beating and killing of Khalid Mohammed Said, as photos of his disfigured body were posted over the Internet and went viral. An influential Facebook group called *We are All Khalid Said* was set up, and the killing provided a tangible focus for solidarity among young Egyptians.

This led to protests with hundreds of thousands of young Egyptians taking to the streets and gathering in Tahrir Square in Cairo. They demanded an end to police brutality as well as the end of the 30-year reign of President Hosni Mubarak. The authorities reacted swiftly in closing the Internet in Egypt, but this act of censorship failed to stop the protests. Social media played an important role in mobilizing protests and in influencing the outcome of the revolution.

48.3   The Tweet

Twitter is a social communication tool that allows people to broadcast short messages. It is often described as the *SMS of the Internet*, and it is an online social media and micro-blogging site that allows its users to send and receive short 140-character messages called *tweets*. The restriction to 140 characters (recently doubled to 280 characters) is to allow Twitter to be used on non-smartphone mobile devices. Twitter has over 300 million active users, and it is one of the most visited web sites in the world. Users may access Twitter through its web site interface, a mobile device app, or SMS.

Jack Dorsey (Fig. 48.3) and others founded the company in 2006. Dorsey introduced the idea of an individual using an SMS service to communicate with a small group while he was still a student at New York University. The word *twitter* was the name chosen for this new service, and its definition as *a short burst of information* and *chirps from birds* was highly appropriate.

Twitter messages are often about friends telling one another about their day, what they are doing, where they are, and what they are thinking and doing, and Twitter has transformed the world of media, politics, and business. It is possible to include links to web pages and other media as a tweet. News such as natural disasters, sports results, and so on are often reported first by Twitter. The site has transformed political communication in a major way, as it allows politicians and their followers to debate and exchange political opinions. It allows celebrities to engage and stay in contact with their fans, and it provides a new way for businesses to advertise its brands to its target audience.

As a Twitter user, you select which other people who you wish to follow, and when you follow someone, their tweets show up your *Twitter stream* (a list of

Fig. 48.3 Jack Dorsey at
the 2012 Time 100 Gala

tweets). Similarly, anyone that chooses to follow you will see your tweets in their stream.

A *hashtag* is an easy way to find all the tweets about a given topic of interest, and it may be used even if you are not following the people who are tweeting. It also allows you to contribute to the topic that is of interest. A hashtag consists of a short word or acronym preceded by the hash sign (#), and conferences, hot topics, and so on often have a hashtag.

A word or topic that is tagged at a greater rate than other hashtags is said to be a *trending topic*, and this is often the result of an event that prompts people to discuss the topic. Trending may also result from the deliberate action of certain groups (e.g., in the entertainment industry) to raise the profile of a musician or celebrity (and to market their work).

Twitter has become an effective way to for an organization to communicate its latest news, and its effectiveness as a communication tool is dependent on the number of followers of the organization. It allows the organization to determine what people are saying about it, as well as their positive or negative experience in interacting with it. This enables the organization to directly interact with its followers, and it is a powerful way to engage and to make people feel heard. It allows the organization to respond to any negative feedback, sensitively and appropriately.

Twitter has experienced rapid growth from 400,000 tweets posted per quarter in 2007 to 500 million tweets per day in 2017. Twitter's usage spikes during important events such as major sporting events, natural disasters, the death of a celebrity, and so on. For such events, there may be over 100,000 tweets per second.

Twitter's main source of revenue is advertisements through *promoted tweets* that appear in a user's timeline (Twitter stream). The first promoted tweets appeared from late 2011, and the use of a tweet for advertisement was ingenious. It helped to make the advertisement feel like part of Twitter, and it meant that an advertisement could go anywhere that a tweet could go. Advertisers are only charged when the user follows the links or re-tweets the original advertisements. Further, the use of tweets for advertisement meant that the transition to mobile was easy, and today about 80% of Twitter use is on mobile devices. For more detailed information on Twitter, see (Schaefer 2014).

48.4 Social Media and Fake News

Fake news is the systematic spreading of misleading or false information in traditional print or online social media, with the intention of misleading or damaging another person or institution. It can negatively affect individuals and lead to violence or hate against minority ethnic groups in a country. The popularity of social media sites such as Facebook and Twitter has contributed to the spread of fake news, and this poses threats to twenty-first-century democracy. Fake news may be spread by individuals, organizations, and hostile states, and it consists of news that has no basis in fact but which is presented as being factually correct.

Fake news in the form of propaganda has been around for centuries, where such news is generally published for political reasons. Military leaders have often embellished their bravery and results in battle throughout history (e.g., Ramses II's description of the Battle of Kadesh in the thirteenth century B.C. paints a very positive but factually inaccurate account of the battle).

Following the invention of the printing press in the fifteenth century, news publications became popular, and over time, fake news stories appeared in the print media. Fake news played an important role in propaganda during the first and second world wars, with radio broadcasts and printed material used to persuade the public at home as well as discouraging enemy troops. Today, modern society is highly dependent on accurate information in the print, radio, television, and online media. The effectiveness of fake news increases when the stories spread widely (as often occurs in social media) and where users interact with and rely on these stories rather than on traditional news media.

Fake news played a key role in the 2016 presidential election in the United States, which led to the election of Donald Trump. Most of the fake election news stories in the last 3 months of the campaign were anti-Clinton, but it is difficult to determine the extent to which this influenced the outcome of the election. Trump and his supporters seem to use the word "fake news" to refer to the mainstream media that is opposed to him and his policies.

It is important when considering the accuracy of an article to consider the source of the news (e.g., is it written by a reputable news organization such as the BBC or Reuters?), as well as considering the authenticity of its authors and the supporting sources. Fake news is a deeply disturbing Internet trend that needs to be resolved if technology is to serve humanity. Modern technology has provided many benefits to modern society, but its abuse needs to be managed.

Fake news is a dangerous trend in society, as false news can spread easily due to the speed and accessibility of modern technology. It allows individuals to be misled and negatively influenced. Online social media sites such as Facebook and Twitter have a responsibility to develop appropriate solutions to address this serious problem.

Chapter 49
Software Inspection Methodology

The approach to software development in the 1950s and 1960s has been described as the "Mongolian Hordes Approach" by Fred Brooks. The "method" or lack of method was characterized by the philosophy:

> The completed code will always be full of defects.
> > The coding should be finished quickly to correct these defects.
> > Design as you code approach.

This philosophy accepted defeat in software development and suggested that irrespective of a solid engineering approach, the completed software would always contain lots of defects and that it therefore made sense to code as quickly as possible and to then identify the defects that would be present, to correct them as soon as possible.

The objective of software inspections is to build quality into the software product, rather than adding quality later. There is clear evidence that the cost of correction of a defect increases the later that it is detected, and it is therefore more cost-effective to build quality rather than adding it later in the development cycle. Software inspections are an effective way of doing this.

Michael Fagan of Michael Fagan Associates is the creator of the Fagan Inspection and Defect Free Process (Fagan 1976). He created the inspection process while he was a Development Manager at IBM in the 1970s. This process helps organizations to improve software quality, reduce cycle time, reduce costs, and improve productivity.

Michael Fagan (Fig. 49.1) was one of the founder members of the IBM Quality Institute in 1981. He received an individual corporate achievement award from IBM for creating the Fagan inspection process and for promoting its use throughout IBM.

He founded Michael Fagan Associates in 1989 to teach others how to use this inspection process effectively and to improve software quality in organizations throughout the world.

© Springer Nature Switzerland AG 2018

G. O'Regan, *The Innovation in Computing Companion*,

https://doi.org/10.1007/978-3-030-02619-6_49

Fig. 49.1 Michael Fagan

49.1 Fagan Inspection Process

Software inspections play an important role in identifying faults in the software development life cycle and in building quality into the software product. The cost of correction of a defect increases the later that it is detected in the development cycle. Consequently, there is an economic argument to identify defects as early as possible, as it is more cost-effective to build quality into the software product rather than adding it later in the development life cycle.

The Fagan inspection process identifies and removes defects in the work products. It stipulates that requirement documents, design documents, source code, and test plans all be formally inspected by experts who are independent of the author. Further, the work product should be examined from different viewpoints (e.g., requirements, design, and test). The quality of the software product is only as good as the quality at the end of each phase, and software inspections assist in ensuring that quality has been built into each phase and that the final product is fit for purpose.

The seven-step Fagan inspection process includes planning, overview, preparation, inspection, process improvement, rework, and follow-up activity. Its objectives are to identify and remove errors in the work products and to identify and correct any systemic defects in the processes used to create the work products[1].

There are various roles defined in the inspection process, including the *moderator*, who chairs the inspection; the *reader*, who paraphrases the deliverable and gives an independent viewpoint; the *author*, who is the creator of the deliverable; and the *tester*, who is concerned with the testing viewpoint. The inspection process

---

[1] A defective process may lead to downstream defects in the work products.

will consider the correctness of the design with respect to the requirements and whether the source code is correct with respect to the design.

The goal is to identify as many defects as possible and to confirm the correctness of the deliverable. Inspection data are recorded and may be used to assess how effective the project is in detecting and preventing defects.

The moderator records the defects identified during the inspection, and the defects are classified according to their type and severity. Mature organizations typically enter defects into an inspection database to allow metrics to be generated and to enable analysis to be performed. The seven stages in the inspection process are summarized in Table 49.1.

There is clear evidence that software inspections have positive impacts on productivity, quality, time to market, and customer satisfaction. For example, IBM Houston employed software inspections for the Space Shuttle missions, and the

Table 49.1 Overview of Fagan inspection process

Activity	Role	Description
Planning	Moderator	Identify inspectors and roles
		Verify material is ready for inspection
		Distribute inspection material
		Book a room for the inspection
Overview	Author	Brief participants on material
		Give background information
Preparation	Inspectors	Prepare for the meeting and role to be performed
		Checklist may be employed
		Read through the deliverable and mark up issues/questions
Meeting	Inspectors	The moderator will cancel the inspection if inadequate preparation done
		Time limit set for inspection moderator keeps meeting focused
		The inspectors perform their roles
		Emphasis on finding defects not solutions
		Defects are recorded and classified
		Author responds to any questions
		The duration of the meeting is recorded
		An inspection outcome is agreed
Process improvement	Inspectors	Continuous improvement of development and inspection process
		The causes of major defects are recorded
		A root cause analysis is performed to identify any systemic defect with the software development process or inspection process
		Recommendations are made to the process improvement team
Rework	Author	The author corrects the defects and carries out any necessary investigations
Follow-up	Moderator/ author	The moderator verifies that the author has resolved the defects and investigations

results showed that 85% of the defects were found by inspections and 15% were found by testing. There were no defects found in the space missions. This project includes about 2 million lines of computer software. IBM, North Harbor, in the United Kingdom quoted a 9% increase in productivity with 93% of defects found by software inspections.

Software inspections also play an important role in educating and training new employees about the product and the standards and procedures to be followed. Sharing knowledge reduces dependencies on key employees. Higher-quality software leads to improved productivity, as less time is devoted to reworking the defective product.

The cost of correction of a defect increases the later it is identified in the lifecycle. Boehm (Boehm 1981) states that a requirement defect identified by the customer is over 40 times more expensive to correct than if it were detected in the requirements phase. The cost of a requirement defect detected at the customer site includes the cost of correcting the requirements and the cost of design, coding, unit testing, system testing, regression testing, and so on. It is more economical to detect and fix the defect in phase.

There are other software inspection methodologies such as Gilb's approach to software inspections (Gilb and Graham 1994).

Chapter 50
Software Life Cycles

The processes employed to develop high-quality software generally include processes for project management, requirement analysis and specification, design and development, peer reviews and testing, selection and management of suppliers, configuration management, audits, and customer support and maintenance.

The software development process has an associated life cycle that consists of various phases. There are several well-known life cycles employed such as the waterfall model (Royce 1970), the spiral model (Boehm 1988), the Rational Unified Process (Jacaobson et al. 1999), and the Agile methodology which was discussed in Chap. 4. The choice of the software development life cycle is determined from the needs of the project.

The waterfall model[1] (Fig. 50.1) is a linear and sequential development method, and it was described by Royce in 1970. It starts with requirement gathering and definition and is followed by the system specification, the design and implementation of the software, and comprehensive testing. The testing generally includes unit, system, and user acceptance testing.

It is employed for projects where the requirements can be determined early in the project life cycle or are known in advance (in practice this is difficult to do). It is also called the "V" life cycle model, with the left-hand side of the "V" detailing requirements, specification, design, and coding and the right-hand side detailing unit tests, integration tests, system tests, and acceptance testing. Each phase has entry and exit criteria that must be satisfied before the next phase commences. The waterfall model allows for managerial control, and a schedule can be created with deadlines for each phase in development. There are several variations to the waterfall model.

Many companies employ a set of templates to enable the activities in various phases to be consistently performed. Templates may be employed for project planning and reporting, requirement definition, design, testing, and so on. The templates may be based on the IEEE standards or on industrial best practice.

---

[1] We are treating the waterfall model as identical to the V model.

© Springer Nature Switzerland AG 2018
G. O'Regan, *The Innovation in Computing Companion*,
https://doi.org/10.1007/978-3-030-02619-6_50

Fig. 50.1 Waterfall V life cycle model

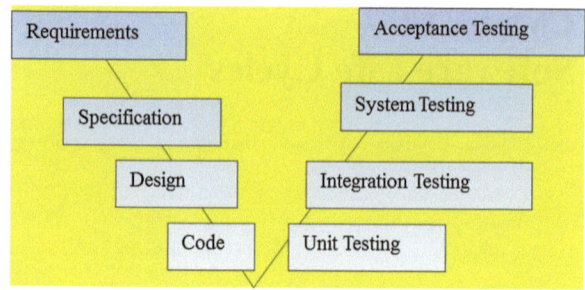

50.1   Spiral Life Cycles

The spiral model (Fig. 50.2) was developed by Barry Boehm in the mid-1980s and is useful for a project in which the requirements are not fully known at project initiation or where the requirements evolve as a part of the development life cycle. The development proceeds in spirals, where each spiral typically involves objectives and an analysis of the risks, updates to the requirements, design, code, testing, and a user review of the iteration or spiral.

The spiral is, in effect, a reusable prototype with the business analysts and the customer reviewing the current iteration and providing feedback to the development team. The feedback is analyzed and used to plan the next iteration. This approach is often used in joint application development, where the usability and look and feel of the application are a key concern. This is important in web-based development and in the development of a graphical user interface (GUI). The implementation of part of the system helps in gaining a better understanding of the requirements of the system, and this feeds into subsequent development cycle. The process repeats until the requirements and the software product are fully complete.

There are several variations of the spiral model including Rapid Application Development (RAD), Joint Application Development (JAD) models, and the Dynamic Systems Development Method (DSDM) model. Agile methods employ sprints (or iterations) of 2-week duration to implement several user stories (see Chap. 4).

There are other life cycle models, for example, the iterative development process that combines the waterfall and spiral life cycle model. The Cleanroom approach includes a phase for formal specification, and its approach to software testing is based on the predicted usage of the software. The Rational Unified Process is discussed in the next section.

50.2   Rational Unified Process

The *Rational Unified Process* (Jacaobson et al. 1999) was developed at the Rational Corporation (now part of IBM). It uses the Unified Modeling Language (UML) as a tool for specification and design, where UML is a visual modeling language.

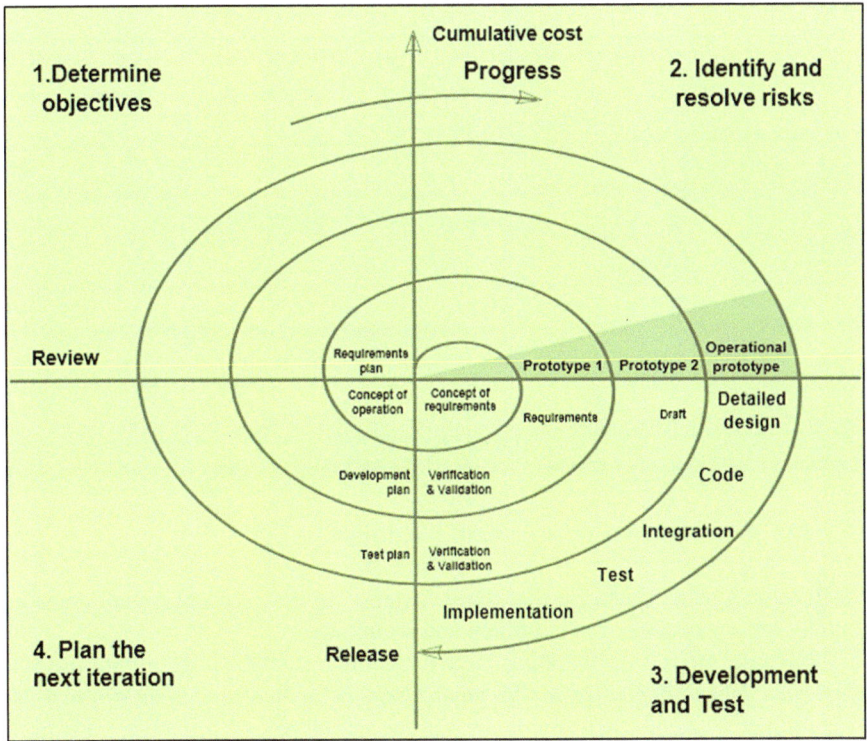

Fig. 50.2 Spiral life cycle model. Public Domain

James Rumbaugh, Grady Booch, and Ivar Jacobson developed UML, and it facilitates the understanding of the architecture and complexity of the system. It provides a means of specifying, constructing, and documenting the object-oriented system.

RUP is *use case driven*, *architecture centric*, *iterative*, and *incremental* and includes cycles, phases, workflows, risk mitigation, quality control, project management, and configuration control (Fig. 50.3). Software projects may be complex, and there are risks that requirements may be incomplete or that the interpretation of a requirement may differ between the customer and the project team.

Requirements are gathered as use cases, and the *use cases describe the functional requirements from the point of view of the user of the system*. They describe what the system will do at a high level and ensure that there is an appropriate focus on the user when defining the scope of the project. *Use cases also drive the development process*, as the developers create a series of design and implementation models that realize the use cases. The developers review each successive model for conformance to the use case model, and the test team verifies that the implementation correctly implements the use cases.

The software architecture concept embodies the most significant static and dynamic aspects of the system. The architecture grows out of the use cases and

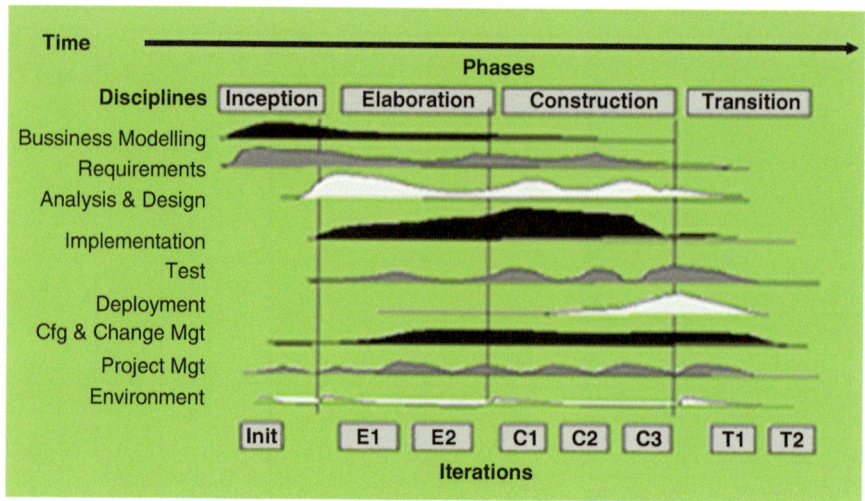

Fig. 50.3 Phases and workflows in Rational Unified Process

factors such as the platform that the software is to run on, deployment consider-ations, legacy systems, and nonfunctional requirements.

RUP decomposes the work in a large project into smaller slices or mini-projects, and *each mini-project is an iteration that results in an increment to the product*. The iteration consists of one or more steps in the workflow and generally leads to the growth of the product. If there is a need to repeat an iteration, then all that is lost is the misdirected effort of one iteration, rather than the entire product. That is, RUP is a way to mitigate risk in software engineering. For further information on software life cycles, see (O'Regan 2017a).

Chapter 51
The System/360 Revolution

The IBM System/360[1] was a family of mainframe computers designed and developed by IBM. It had a major impact on and it placed IBM on the road to dominate the computer field for the next 20 years, up to the development of home and personal computers in the early 1980s.

It was the beginning of an era of computer compatibility, where for the machines across a product line could work with each other. It meant that IBM customers could start off with a low specification member of the 360 family, and upgrade over time to a more powerful member. This allowed the customer to choose the appropriate model to meet its current needs, and then as its needs evolved over time, it could upgrade to a more powerful member. It was a massive $5 billion investment (*bet the business gamble*) by Thomas Watson Jr., and it moved IBM from its traditional business and product lines into the unknown with the gamble that the future would be the IBM System/360.

Thomas Watson Jr.[2] announced the System/360 in 1964, and the revolutionary announcement changed business and the world of computing forever. The System/360 replaced all five of IBM's computer product lines with one strictly compatible family. It used a new computer architecture that employed hybrid integrated circuit technology, and it pioneered the 8-bit byte, which remains in use today.

The System/360 included a multiprogramming disk-based operating system, which was called OS/360. It included free software packages such as compilers for several programming languages, as well as packages for communication network capabilities (Pugh 2009).

It was an extremely successful product line for IBM, with orders rapidly exceeding forecasts. Its success vastly exceeded IBM's expectations, with over a thousand

---

[1] The number "360" (the number of degrees in a circle) was chosen to represent the ability of each computer to handle all types of applications.

[2] Thomas Watson Jr. later stated, "The System/360 was the biggest, riskiest decision that I ever made, and I agonised about it for weeks, but deep down I believed that there was nothing that IBM couldn't do."

© Springer Nature Switzerland AG 2018
G. O'Regan, *The Innovation in Computing Companion*,
https://doi.org/10.1007/978-3-030-02619-6_51

orders placed in the first 4 weeks after the announcement. The popularity of the System/360 made it difficult for IBM competitors (such as Burroughs, Honeywell and Sperry-Rand) to compete against IBM in the general-purpose computer market.

Monthly rental prices ranged from under $3,000 per month for the most basic system to over $100,000 per month for a large multisystem. The purchase cost ranged from $130,000 for a basic system to over $5 million for a large system. In 1989, 25 years after the announcement of the System/360, products based on the System/360 architecture and its extensions still accounted for over 50% of IBM revenue.

51.1   Background to the System/360

Thomas Watson Jr., the son of Thomas Watson Sr. (the first president of IBM), became the president of IBM in 1952. He recognized that computers would play a key role for business in the years ahead and that the future of IBM was in the computer business and not in tabulators. It was clear to him that IBM needed to change, and he played a key role in transforming the company to become the world leader in the computer industry.

IBM was already a successful computer company in the 1950s. It introduced its first large computer (the IBM 701) based on vacuum tubes in 1952, the IBM 650 (Magnetic Drum Calculator) in 1954, and the IBM 704 data processing system computer in 1954. It played a key role in the development of the computers for the SAGE air defense system in the United States. IBM had become the market leader with a large growth in its revenue and earnings, and it employed over 100,000 people around the world.

However, within IBM, there were concerns that the company had reached a plateau and that competitors were launching alternative products to IBM. The origins of the System/360 go back to the late 1950s and Watson's determination to transform IBM to position it for future success. IBM was supporting five different product lines by 1959, and it was becoming a major challenge to train staff to service and maintain software to support so many different computer products.

There were major problems with incompatibility between different hardware and software among the different computer vendors, as well as incompatibility among IBM's own products. IBM had an existing product line of several computers, each excellent, but all with incompatible architectures. It meant that customers who wished to upgrade from their existing system to a larger system had to invest in a totally new system, with new printers, new storage devices, and new software (often totally rewritten for the new machine).

It was clear to Watson and other senior IBM executives that there was a need to develop a totally cohesive product line so that computers produced at different IBM facilities would be compatible with one another.

IBM set up a corporate-wide task group to establish an overall IBM plan for its future products. The task group had the acronym SPREAD (System Programming, Research, Engineering and Design), and it completed its final report in late 1961. It made a series of recommendations such that there should be five processors spanning a 200-fold range in performance. IBM made the brave decision in 1962 to replace the company's entire product line of computers and to build a new family of compatible machines.

It would mean that code written for the smallest member of the family would be upwardly compatible with each of the processors in the family. Further, the various peripherals such as printers and storage devices would be compatible across the family. It was an incredibly brave decision, and Fortune Magazine later described it as "IBM's five-billion-dollar gamble."

51.2   The IBM System/360

Thomas Watson announced the new System 360 to the world at a press conference in 1964 and said:

> The System/360 represents a sharp departure from concepts of the past in designing and building computers. It is the product of an international effort in IBM's laboratories and plants, and is the first time IBM has redesigned the basic internal architecture of its computers in a decade. The result will be more computer productivity at lower cost than ever before. This is the beginning of a new generation – – not only of computers – – but of their application in business, science and government.

The IBM System/360 (Fig. 51.1) was a family of small to large computers, and the concept of a *family of computers* was a paradigm shift away from the traditional *one size fits all* philosophy of the computer industry, as, up until then, every computer model was designed independently.

The family of computers ranged from minicomputers with 24 KB of memory, to supercomputers for US missile defense systems. However, all these computers employed the same user instruction set, and the main difference was that for the larger computers some machine instructions were implemented with hardware, whereas the smaller machines used microcode.

The System/360 architecture allowed customers to commence with a low-cost computer model and to upgrade over time to a larger system to meet their evolving needs. The fact that the same instruction set was employed meant that the time and expense of rewriting software were avoided.

Gene Amdahl (Fig. 51.2) was the chief architect for the System/360, and Fred Brooks[3] was the project manager (Fig. 51.3). The System/360 family was introduced in 1964, and the IBM chairman, Thomas Watson Jr., called it the most important product announcement in the company's history.

---

[3] Fred Brooks wrote an influential paper "The Mythical Man Month" based on his experience as project manager for the System 360 project.

Fig 51.1 IBM System/360. (Courtesy of Courtesy of International Business Machines Corporation, © International Business Machines Corporation)

Fig 51.2 Gene Amdahl.
Creative Commons

Fig 51.3 Fred Brooks. (Photo courtesy of Dan Sears)

The IBM 360 family offered a choice of 5 processors and 19 combinations of power, speed, and memory. There were 14 models in the family. It was successful in achieving strict compatibility in the family of computers, and the project introduced several new industry standards including 8-bit bytes.

A customer could start with a small member of the System/360 family and upgrade over time into a larger computer in the family. This made computers more affordable for businesses, and it stimulated growth in computer use. The System/360 was used extensively in the Apollo program to place man on the moon, where IBM computers and personnel were essential to the success of the project. IBM invested over \$5 billion in the design and development of the System/360, but the gamble paid off and it was a very successful product line.

Gene Amdahl was appointed an IBM fellow in 1965 in recognition of his contribution to IBM, and he was appointed director of IBM's Advanced Computing Systems (ACS) Laboratory in California and given freedom to pursue his own research projects. He later left IBM following disagreements on future computer development, and he formed Amdahl Corporation, which later became a major competitor to IBM in the mainframe market.

Fred Brooks was the project manager for the System/360 project, which involved 5000 man-years of effort at IBM. Brooks recorded his experience as project manager in a famous project management book titled *The Mythical Man-Month* (Brooks 1975). This book which appeared in 1975 considered the challenge of delivering a major project (of which software is a key constituent) on time, on budget and with the right quality. Brooks described it as "my belated answer to Tom Watson's probing question as to why programming is hard to manage."

For a more detailed account of the System/360 revolution, see the excellent IBM article "The 360 Revolution" by Chuck Boyer (Boyer 2004).

Chapter 52
Transistor

The early computers were large bulky machines taking up the size of a large room (e.g., see Chap. 23). They contained thousands of vacuum tubes,[1] and these tubes consumed large amounts of power and generated a vast quantity of heat. This led to problems with the reliability of the early computers, as several tubes burned out each day. This meant that machines were often nonfunctional for parts of the day, until the defective tube was identified and replaced.

There was therefore a need to find a better solution to vacuum tubes, and William Shockley set up the solid physics research group at Bell Labs after the Second World War. His group included John Bardeen and Walter Brattain, and Shockley's goal was to find a solid-state alternative to the existing glass-based vacuum tubes. Bell Lab's invention of the transistor revolutionized the field of electronics.

The *transistor* is a fundamental building block in modern electronic systems. It was smaller, cheaper, and more reliable than the vacuum tubes that were used in the early computers. It is a three-terminal, solid-state electronic device, and it can control electric current or voltage between two of the terminals by applying an electric current or voltage to the third terminal. The three-terminal transistor enables an electric switch to be made which can be controlled by another electrical switch. Complicated logic circuits may be built up by cascading these switches (switches that control switches that control switches and so on).

These logic circuits may be built very compactly on a silicon chip with a density of millions of transistors per square centimeter. The switches may be turned on and off very rapidly (e.g., every 0.000000001 second). These electronic chips are at the heart of modern electron devices.

Shockley was involved in radar research and anti-submarine operation research during the Second World War, and after the war he led the Bell Lab research group to find a solid-state alternative to the glass-based vacuum tubes. Shockley, Bardeen,

---

[1] ENIAC contained over 18,000 vacuum tubes, and the AN/FSQ-7 computer used in the SAGE air defense system contained 55,000 vacuum tubes.

© Springer Nature Switzerland AG 2018
G. O'Regan, *The Innovation in Computing Companion*,
https://doi.org/10.1007/978-3-030-02619-6_52

and Brattain were awarded the Nobel Prize in physics in 1956 in recognition of their invention of the transistor (Fig. 52.1).

Bardeen and Brattain succeeded in creating a point contact transistor in 1947 independently of Shockley who was working on a junction-based transistor. Shockley believed that the point contact transistor would not be commercially viable, and his junction point transistor was announced in 1951. The junction point transistor soon eclipsed the point contact transistor and became dominant in the market place.

Shockley (Fig. 52.2) published a book on semiconductors *Electrons and Holes in Semiconductors with applications to transistor electronics* in 1950 (Shockley 1950). This was the first textbook for scientists and engineers, who were interested in learning about the new field of transistors and semiconductors. He resigned from Bell Labs in 1955 and formed Shockley Laboratory for Semiconductors (part of Beckman Instruments) at Mountain View in California. This company played an important role in the development of transistors and semiconductors, and several of its staff later formed semiconductor companies in the Silicon Valley area.

Shockley was the director of the company, but his domineering management style alienated several of his employees. This led to the resignation of eight key researchers in 1957 following his decision not to continue research into silicon-based semiconductors. This gang of eight went on to form Fairchild Semiconductors and other companies in the Silicon Valley area in the following years. They included Gordon Moore and Robert Noyce, who founded Intel in 1968. Other Fairchild employees formed National Semiconductors and Advanced Micro Devices.

Shockley Semiconductors and these new companies formed the nucleus of what became Silicon Valley. The second generation of computers used transistors instead of vacuum tubes, and next we discuss a selection of these.

Fig. 52.1 Replica of Transistor. Public Domain

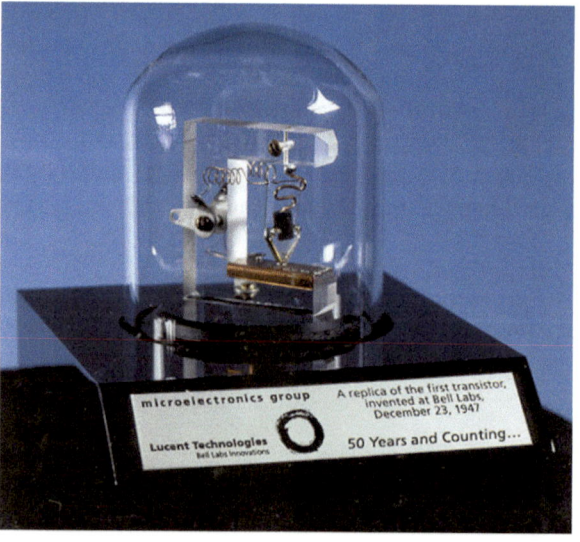

Fig. 52.2 William
Shockley. Creative
commons

52.1   Early Transistor Computers

The University of Manchester Experimental Transistor Computer was one of the first transistor computers when it was introduced in 1953.[2] The prototype machine used 92-point contact transistors and had a 48-bit word size, whereas the full-scale version used 200-point contact transistors. There were serious problems with the reliability of the point contact transistors, which meant that the machine itself was unreliable. Metropolitan-Vickers (a Manchester company) adapted the design and changed the circuits to use the more reliable junction-based transistors, and it created a full-scale version called the Metrovick 950 in 1956.

Other early transistor computers include the TRADIC designed and developed by Bell Labs in early 1954. This machine also used some vacuum tubes. The Harwell CADET was an early fully transistorized machine when it appeared in early 1955. The IBM 608 was the first IBM product to use transistor circuits instead of vacuum tubes. The prototype of this product appeared in 1955, and the fully transistorized calculator was introduced in late 1957. It contained 3000 germanium transistors. The Burroughs SM-65-Atlas ICBM was an early transistorized computer, which appeared in 1957.

---

[2] It was not a fully transistorised computer in that it employed a small number of vacuum tubes in its clock generator.

The IBM 7090 was one of the earliest commercial computers with transistor logic, and it was introduced in 1958. It was designed for large-scale scientific applications, and it was over 13 times faster than the older vacuum tube IBM 701. It used 36-bit words, had an address-space of 32,768 words, and could perform 229,000 calculations per second. It was used by the US Air Force to provide an early warning system for missiles and by NASA to control space flights. It cost approximately $3 million, but it could be rented for over $60 K per month.

Chapter 53
UNIX Operating System

Ken Thompson, Dennis Ritchie, and others designed and developed the UNIX operating system at Bell Labs in the early 1970s. UNIX is a multitasking and multiuser operating system that was written almost entirely in the C programming language (which was designed by Denis Ritchie at Bell Labs). It arose out of work done by Massachusetts Institute of Technology, General Electric, and Bell Labs on the development of a general timesharing operating system called *Multics*. This was a large and complex operating system that was intended to be all things to all people, and it was designed to be used on General Electric's GE-645 mainframe computer.

Bell Labs decided to withdraw from the Multics project in 1969, as they believed that it would be a large expensive system. They decided instead to use General Electric's GECOS operating system. However, several of the Bell Lab researchers (led by Ken Thompson) decided to continue the work on a less ambitious smaller-scale operating system, and the name "UNIX" was coined by Brian Kernighan (as a pun on Multics).

The first version of the operating system was written on a Digital PDP-7 mini-computer in assembly language, and it included a file system, a process control mechanism, file utilities, and a command interpreter. Dennis Ritchie joined the project, and he helped in rewriting UNIX in the C programming language in 1973 for the PDP-11 computer (which had been introduced by DEC in 1970).

Thompson and Ritchie received the ACM Turing Award for their design and development of UNIX in 1983 (Fig. 14.1). AT&T made the UNIX operating system available to universities for a nominal fee, and the source code was also distributed. This led to further development of the operating system, and Microsoft introduced XENIX, a commercial version of UNIX, in 1980. It included extra features for hardware error recovery and automatic file repair after crashes or power failure.

The use of C helped to make UNIX more portable, and it became a widely used operating system. Universities and the US government used it initially, but it later became popular in industry. It is a powerful and flexible operating system, and it is used on a variety of machines from micros to supercomputers. It is designed to

© Springer Nature Switzerland AG 2018
G. O'Regan, *The Innovation in Computing Companion*,
https://doi.org/10.1007/978-3-030-02619-6_53

allow several programmers to access the computer at the same time, and to share its resources, and it offers powerful real-time sharing of resources.

UNIX has been implemented on a wide variety of supercomputers, mainframes, minicomputers, and personal computers. The UNIX system provides software tools that support program development environments.

It includes features such as *multitasking* which allows the computer to do several things at once, *multiuser* capability which allows several users to use the computer at the same time, and *portability* of the operating system which allows it to be used on several computer platforms with minimal changes to the code. It includes a collection of tools and applications. There are three levels of the UNIX system: *kernel*, *shell*, and *tools and applications*.

The kernel is the central part of the UNIX operating system, and it provides system services to application programs. This includes services for process management, memory management, and input/output management. UNIX manages many concurrent processes.

The UNIX shell is a command interpreter that acts as the interface between the user and the operating system. There are several popular shells for UNIX including the Bourne shell and Korn shell. UNIX uses a hierarchical file system with the root node at its origin and with each directory entry containing files and other directories.

UNIX was initially used by universities and the US government, but it later became popular in industry. For a more detailed account of the UNIX operating system, see (Deitel 1990, Robbins 2005).

53.1 UNIX Shell

The UNIX shell acts as the interface between the user and the operating system. It is a command interpreter that reads lines typed by the user and then executes the required system features. The UNIX shell is an application program and is not part of the kernel of the operating system. UNIX supports several different shells, and the three most popular shells are the *Bourne shell* (program file *sh*), the *Berkeley C shell* (program file *csh*), and the *Korn shell* (program file *ksh*).

The shell issues a prompt such as "$" or "%" to indicate that it is ready for the next command line, and a command line consists of a command name followed by a list of arguments separated by blanks. The command languages interpreted by the various shells are substantial.

The shell breaks up the command line into its components, and a *fork* is performed creating a child process containing the shell. The file specified in the command is loaded and the arguments made available to it. The child executes the command, and the parent process waits for the child to terminate with an exit call. The shell process then displays its prompt to indicate that it is ready for the next command.

The shell gives each program it executes three open files, and these are *standard input* (usually the keyboard), the *standard output* (usually the display), and the *standard error* (usually the display).

53.2   UNIX Kernel

The kernel is the central part of the UNIX operating system, and it provides services to the application programs (including the shell) and hides the hardware from the user. The kernel provides process management, memory management, I/O management, and timer management.

UNIX manages many concurrent processes, and each process has its own address space. Some system calls are privileged and may be executed only by privileged processes. The *superuser* (system administrator) creates privileged processes, whereas processes created by other users are generally unprivileged. UNIX creates and destroys processes quite frequently, and the process environment consists of a code region, a data region, a stack region, and various data structures.

The *fork* system call creates a new *child* process, which is a copy of the parent process. Processes may be terminated voluntarily by the exit call or because of illegal actions generated by the user.

UNIX uses a hierarchical file system with the root node at its origin, and file names appear in directories that are UNIX files. Each UNIX file has three sets of permission bits associated with it, and these bits control *read access*, *write access*, and *execute access*. A file may be identified by an *absolute path name* or a *relative path name*. For example, the absolute path name is interpreted as the path

/usr/gor/books/inv.txt

starting at the root and then going through the *usr* directory, the *gor* directory, and the *books* directory. Any file without a slash is interpreted as being relative to the user's *current directory*.

Processes are scheduled according to their priorities. Concurrent processes may communicate with each other using *interprocess communication*. This may include approaches such as pipes, messages, and shared memory. A *pipe* is a unidirectional path over which processes may send streams of data to other processes, and it allows the output of one process to be used as the input to another in FIFO order. A *signal* is a software mechanism to indicate the occurrence of an event such as a hardware interrupt.

The input/output is performed on streams of bytes, and there is no concept of records. Any structure in the byte streams is imposed by the application programs. There are several system calls specific to input/output in UNIX including *open*, *close*, *read*, *write*, and *lseek*.

Chapter 54
Von Neumann Architecture

The earliest computers were fixed-program machines that were designed to do a specific task. This proved to be a major limitation as it meant that a complex manual rewiring process was required to enable the machine to solve a different problem.

John von Neumann (Fig. 54.1) gave his name to what has become known as the *von Neumann architecture* that is used in almost all computers. Eckert and Mauchly were working with him on this concept during their work on ENIAC and EDVAC (see Chap. 23), but their names were not included in the published draft report. This was due to their resignation from the University of Pennsylvania to protect their intellectual property on ENIAC and EDVAC and to commercialize their ideas with the formation of their own computer company.

The von Neumann draft report (*First Draft of a Report on the EDVAC*) was controversial, as it was treated as a legal publication. This meant that it was a public disclosure of the invention of EDVAC and that Mauchly and Eckert could not patent EDVAC, as it was already in the public domain. Further, it was claimed that the stored program concept had evolved out of meetings at the Moore School at the University of Pennsylvania prior to the involvement of von Neumann as a consultant and that credit for this concept should be given to others as well as to von Neumann.

The von Neumann architecture includes a central processing unit which includes the control unit and the arithmetic unit, an input and output unit, and memory. Von Neumann made important contributions to mathematics and early computing, and he created the field of cellular automata and is the inventor of the *merge-sort algorithm* (in which the first and second halves of an array are each sorted recursively and then merged). He also invented the *Monte Carlo* method that allows complicated problems to be approximated using random numbers.

The von Neumann architecture (Fig. 54.2) arose out of work done by von Neumann, Eckert, Mauchly, and others on the design of the EDVAC computer (which was the successor to ENIAC). Von Neumann's draft report on EDVAC described the new architecture (von Neumann 1945), and it remains the fundamental

© Springer Nature Switzerland AG 2018
G. O'Regan, *The Innovation in Computing Companion*,
https://doi.org/10.1007/978-3-030-02619-6_54

Fig. 54.1 John von
Neumann

Fig. 54.2 Von Neumann
architecture

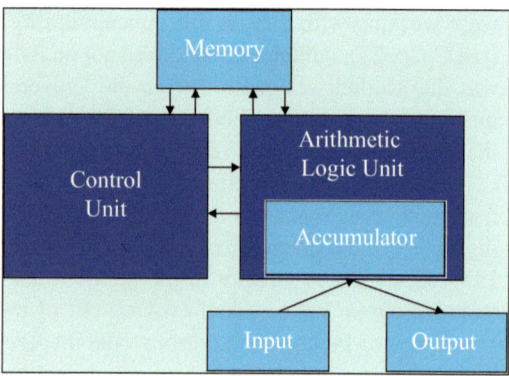

architecture underlying the computers used today, which are general-purpose
machines designed to allow a variety of programs to be run on the machine.

Von Neumann architecture led to the birth of stored program computers, where a
single store is used for both machine instructions and data. Its key components are
described in Table 54.1.

The key approach to building a general-purpose device according to von
Neumann was in its ability of not only to store its data and intermediate results of
computation but also to store the instructions (or commands) for the computation.
The computer instructions can be part of the hardware for specialized machines, but
for general-purpose machines, computer instructions must be as changeable as the
data that is acted upon by the instructions.

Table. 54.1 Von Neumann architecture

Component	Description
Arithmetic unit	The arithmetic unit can perform basic arithmetic operations
Control unit	The control unit contains a built-in set of machine instructions, and the program counter contains the address of the next instruction to be executed
	This instruction is fetched from memory and executed
	This is the basic fetch-and-execute cycle (Fig. 54.3)
Input-output unit	The input and output unit allows the computer to interact with the outside world
Memory	There is a one-dimensional memory that stores all the program instructions and data. These are usually kept in different areas of memory
	The memory may be written to or read from, i.e., it is random-access memory (RAM)
	The program instructions are binary values, and the control unit decodes the binary value to determine the instruction to execute

Fig. 54.3 Fetch-and-execute cycle

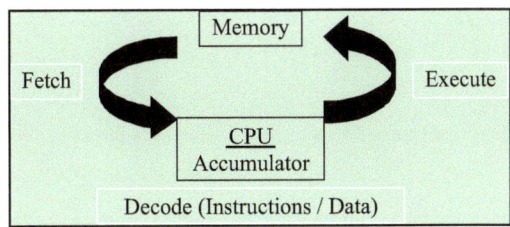

Von Neumann's insight was to recognize that both the machine instructions and data could be stored in the same memory.

The key advantage of the von Neumann architecture was that it was much simpler to reconfigure a computer to perform a different task. All that is required is to enter new machine instructions in memory, rather than physically rewiring a machine (as was required with ENIAC). The limitations of von Neumann architecture include that it is limited to sequential processing rather than parallel processing.

Chapter 55
Wi-Fi Technology

Wi-Fi is a popular short-range wireless broadband technology based on the IEEE 802.11 standard, and it is the wireless equivalent of Ethernet. It uses 2.4–5 gigahertz (GHz) radio frequencies, and it can transfer data at rates of up to a maximum of 600Mbps. The term "Wi-Fi" is a trademark of the Wi-Fi Alliance, and the name was coined by a brand consulting company (Wi-Fi sounds a little like hi-fi). The technology is used for most home networks, business local area networks, and pubic hotspot networks (Fig. 55.1).

Wi-Fi technology allows local area networks (LANs) to operate without cable or physical connections, and it eliminates the need for complex cabling and network switches and connectors. It may be used to connect computers and Wi-Fi-compatible devices to each other, to the Internet, and to the wired network. Wi-Fi-enabled devices can connect to the Internet when they are near areas that have Wi-Fi access called *hotspots*.

Wi-Fi grew out of a technology called WaveLAN designed by AT&T and NCR for wireless cash registers in the late 1980s, and the technology was developed further in the 1990s. It was published as the 802.11 standard in 1997, and it initially provided 2 Mb/s link speeds. The standard evolved over time (e.g., 802.11 g with link speeds of 54 Mb/s) to provide increased performance and features. Wi-Fi began to take off from the early 2000s, especially with high-speed broadband in the home, as it provided an easy way for several computers to share the same broadband link.

Wi-Fi-compatible devices (e.g., personal computers, tablets, smartphones, and printers) within range can all connect to and communicate through a central wireless access point. The *access point* (AP) is a wireless LAN transceiver or base station that can connect one or more wireless devices simultaneously to the Internet. Home networks often use a wireless broadband router as a Wi-Fi access point, whereas public hotspots often use one or more access points inside the coverage area.

Small Wi-Fi radios and antennas are embedded inside smartphones, laptops, and tablets allowing them to function as network clients, and the access points are

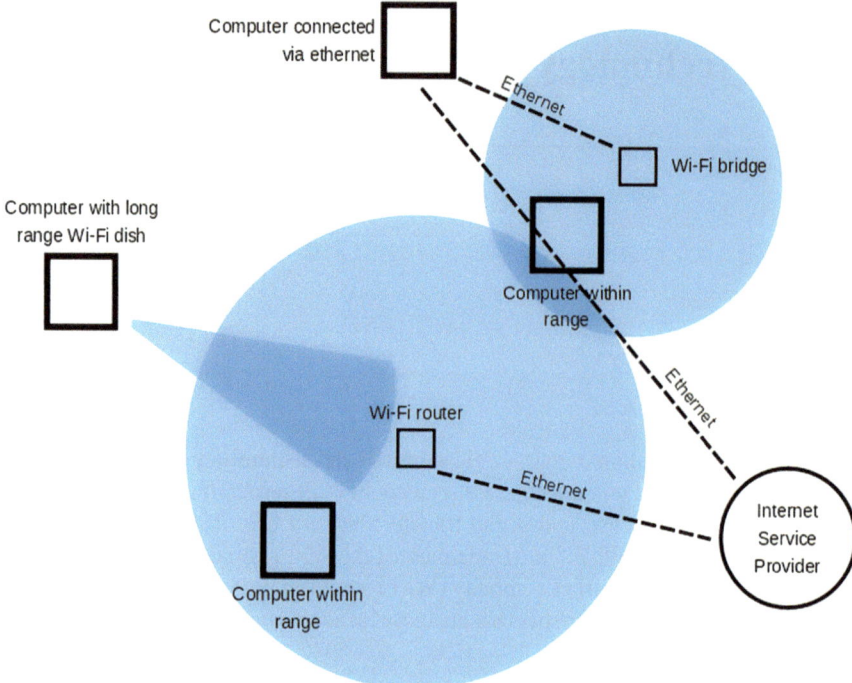

Fig. 55.1 Wi-Fi range diagram. Public domain

configured with network names that the user can identify when scanning the area for available networks.

A Wi-Fi hotspot is a physical location where a person may obtain Internet access via Wi-Fi technology. This is done using a wireless local area network (WLAN) connected to an Internet service provider via a router. Hotspots are places where you can bring your laptop or other mobile device and connect to the Internet. These include venues such as libraries, coffee shops, and schools. Further, many towns and cities have developed and implemented free Wi-Fi networks for both visitors and their residents. There are several million hotspots in use around the world, and while most are free to users, some operate on a pay per service model or on a subscription basis. Many hotspot access points use software for managing user subscriptions and limiting Internet access.

Modern electronic devices can easily determine if there are wireless networks in range, and the user may then initiate a connection to the hotspot's wireless network. The range of a Wi-Fi network varies depending on obstructions that the radio signals encounter between network endpoints. A range of 30 m indoors is typical (and up to 95 m outdoors), but if there are significant obstructions in the radio signal's path, then the range could be much smaller. It is possible to improve the range with specialized antennae.

Wi-Fi networks are more vulnerable to eavesdropping attacks than wired networks such as Ethernet, but security technology has been developed to help to address this (e.g., the Wi-Fi Protected Access (WPA2) encryption).

Many wireless carriers are beginning to offer unlimited mobile data plans, and so many consumers may no longer need to sign into a Wi-Fi network to avoid expensive network charges. However, if the user is on holiday in another country, there is a possibility of roaming charges being applied by the network operator, and Wi-Fi will remain important in these situations. Wi-Fi is likely to continue to be important in homes and office buildings, but it may become less important over time as new technologies overtake it.

55.1 Wi-Fi Security

Wi-Fi hotspots are widely used around the world, with hotels, airports, and cafes offering free Wi-Fi to their customers. This provides convenience to travelers, but there are dangers associated with Wi-Fi including risks of hostile attack and information being stolen or compromised. It is therefore important to understand how these attacks work so that the associated risks may be managed and to ensure that data is kept secure and that sensitive information is protected.

Once a user connects to a public Wi-Fi network, information is sent out into the world, and there are several ways in which this information may be compromised. These may include sniffing attacks, rogue access points, and evil twin attacks (Table 55.1).

Table 55.1 Methods for intercepting data

Method	Description
Sniffing	A sniffing attack does not require sophisticated technical expertise, as a sniffing device examines information passed over the network and captures information such as passwords. Encryption helps to eliminate these attacks, but some older protocols (e.g., WEP) are easy to crack and provide limited protection against hostile attack
Rogue access points	This is a simple form of the *man-in-the-middle* attack where the problem is that you cannot be sure of what you are directly connecting to with a public hotspot. The attacker's laptop is configured to act as an access point (with an innocent name), and everything looks fine to the victim with access provided to the Internet. However, the attacker captures all information, with unencrypted information immediately compromised, and encrypted information may be decoded (depending on the complexity of the encryption method employed)
Evil twin attacks	This is a variant of the rogue access point where the attacker sets up an identical network name to an existing public hotspot and tricks the victim's device into connecting to the evil twin rather than the legitimate hotspot

Chapter 56
Wikipedia

The idea of compiling the world's knowledge into a single location dates back to the ancient library of Alexandria. Ptolemy I, who was a Macedonian general, founded this famous library in the third century B.C. He became the ruler of the Egyptian part of Alexander the Great's empire after Alexander's death in 324 B.C., and the Ptolemaic dynasty ruled Egypt up to the death of Cleopatra in 30 B.C. The library in Alexandria was one of the largest and most important of the ancient world, with most of the books in the library kept as papyrus scrolls. The estimates of the number of books in the collection vary between 40,000 and 400,000 scrolls.

An *encyclopedia* is a compendium of knowledge, and it consists of an extensive summary of many branches of knowledge. It consists of articles organized alphabetically by article name, and a concise description is provided in each article. The earliest encyclopedia (*Naturalis Historia*) dates back to the Roman period, and it was written (in *Latin*) by Pliny the Elder circa 79 A.D. Archbishop Isidore of Seville wrote an encyclopedia (the *Etymologiae*) in the middle ages (c. seventh century A.D.), and this was a compilation of the known learning in the world. Johannes Gutenberg invented the printing press in the fifteenth century during the Renaissance period, and it meant that books no longer needed to be copied by hand (a slow and expensive process). Books could now be mass-produced leading to wider circulation and reading of encyclopedias.

Modern encyclopedias evolved out of dictionaries from the eighteenth century, with Diderot's *Encyclopédie* published in 1751 and the *Encyclopedia Britannia* first published in 1771. These provided a comprehensive set of topics that were discussed in depth, and the 2010-printed version of the *Encyclopedia Britannia* contained 32 volumes and 32,000 pages. The twentieth century saw the appearance of the *Children's Encyclopedia* and specialized encyclopedias in specific fields.

Vannevar Bush outlined his vision of an information management system called the *memex* (memory extender) in a famous essay *As We May Think* (Bush 1945). He envisaged the memex as a device electronically linked to a library that would be able to display books and films. It describes a proto-hypertext computer system where the individual stores all his books, records, and communications and which

© Springer Nature Switzerland AG 2018
G. O'Regan, *The Innovation in Computing Companion*,
https://doi.org/10.1007/978-3-030-02619-6_56

is mechanized so that it may be rapidly consulted and acts as a supplement to his memory. Bush predicted:

> Wholly new forms of encyclopedias will appear, ready made with a mesh of associative trails running through them, ready to be dropped into the memex and there amplified.

Wikipedia is a free open-source Internet encyclopedia that anyone can edit, and Jimmy Wales and Larry Sanger founded this multilingual collaborative encyclopedia in 2001. It is written and edited by a worldwide community of unpaid volunteers, and there is no central organization controlling the editing. This contrasts with other encyclopedias such as *Encyclopedia Britannica*. Wikipedia is a useful educational resource where a user can go to learn about a topic.

It is owned by the nonprofit organization Wikimedia Foundation, and there are over 200 Wikipedia encyclopedias in various languages. The English Wikipedia is the largest of these with over 5 million articles, and there are a total of 40 million articles in various languages. Wikipedia is in the top 10 of the most popular websites, and the Wikipedia foundation handles the servers and legal issues.

Wikipedia grew out of the Nupedia project, which was a free online encyclopedia with articles written by experts and subject to a strict formal review process prior to publication. The Nupedia articles were written by highly qualified volunteers who were experts in their field, and the authors usually had a PhD in their discipline. The articles were subjected to a rigorous peer review to assure their quality.

However, despite its full-time editor-in-chief (Larry Sanger) and a mailing list of editors, the production of content for Nupedia was extremely slow. There were about 12 articles published the first year, and approximately 150 articles were still in draft format. This meant that Nupedia was useless as an encyclopedia, and it was clear that there was a need for a radically new approach so that content could be created much faster (Fig. 56.1).

Fig. 56.1 Wikipedia logo.
Creative Commons

Wikipedia was initially intended to complement Nupedia by providing additional draft articles and ideas for it, and the goal was to create content faster for Nupedia rather than creating a separate online encyclopedia. It was developed as a wiki-style web site, and the project was given the name *Wikipedia* (which came from wiki and encyclopedia), and it was launched on its own domain *wikipedia.com*.

The wiki took off and Wikipedia quickly overtook Nupedia and became a global project. It generated web content in a similar way to GNU and the free open-source software movement, with content created and maintained by a worldwide community of volunteers. The site is not elitist, and anyone may edit a page, and the results show up instantly. This may lead to inaccurate content, and if a bad edit appears, then the community of volunteers may get rid of it by clicking on a revert link.

Wikipedia articles aim to maintain a neutral point of view even on controversial topics. However, consensus can be quite difficult to achieve on sensitive topics, as it may be a challenge to find common ground.

Wikipedia has led to the demise of the printed versions of commercial encyclopedias, as these are unable to compete with a product that is essentially free. Traditional printed encyclopedias (e.g., *Encyclopedia Britannia*) are now historical and are available online only, with Wikipedia being the largest web-based encyclopedia in the world. Wikipedia is owned by the nonprofit organization called the Wikimedia Foundation, and it is funded by donations from its users. The funds are used to pay for the technology to run the organization and to cover the salaries of its staff.

Academics, historians, and journalists have criticized Wikipedia as being an unreliable source of information. They argue that its content is a mixture of truth, half-truths, and incorrect information and that it is especially unreliable when dealing with controversial topics. However, Wikipedia is often a good starting point to learn about a topic, and the accuracy of its articles is constantly improving, as its worldwide community includes experts in many fields. Jimmy Wales inspiring mission for Wikipedia is:

> Imagine a world in which every single person on the planet is given free access to the sum of all human knowledge. That's what we're doing.

56.1 Wikipedia Quality Controls

Wikipedia has processes and structures in place to control the editing of articles (as part of its editorial control), and it uses several approaches to ensure that its content is as accurate and unbiased as possible.

There are thousands of regular editors who vary in knowledge from experts in their field to more casual readers. Anyone who is not a blocked user may visit Wikipedia and edit an article on the site. There are mechanisms for the Wikipedia community to spot bad edits, and a few hundred administrators have the authority to enforce good behavior. There is an arbitration committee that considers situations that remain unresolved, and this committee has the authority to impose sanctions (including a restriction of editing privileges).

The Wikipedia community is largely self-organizing, and while poor information may be added to the site, over time other editors will amend the article until consensus is reached. That is, the approach is in a way like group learning, which leads to quality improvement of the article through successive edits.

Anyone may build a reputation as a competent editor over time, and they may choose to become involved in more specialized roles such as reviewing articles at the request of others or watching newly created articles or existing articles for accuracy.

There are software controls that make it easy for editors to check for acts of vandalism (malicious edits) and to monitor recent changes and to check activity in articles in personal watchlists. Automated software controls (e.g., the program *VandalProof*) allow bad edits to be removed at the click of a button and allow problematic editors to be blocked. These software controls generally allow vandalism to be identified and corrected within minutes.

Wikipedia also has systems in place for article review and improvement, including quality-based peer reviews where editors are invited to comment on an article that they were not involved in writing. The review will consider the readability, quality, and balance of the article, as well as its compliance to Wikipedia policies.

Chapter 57
World Wide Web

The vision of the Internet and World Wide Web goes back to an article written by Vannevar Bush in the 1940s. Bush was an American scientist who worked on submarine detection for the US Navy. He designed and developed the differential analyser (Fig. 1.1), which was a mechanical computer whose function was to evaluate and solve first-order differential equations. It was funded by the Rockefeller Foundation and developed by Bush and others at MIT in the early 1930s.

Bush (Fig. 57.1) became director of the Office of Scientific Research and Development, and he developed a win-win relationship between the US military and universities. He arranged large research funding for the universities to carry out applied research to assist the US military. This allowed the military to benefit from the early exploitation of research results, and it also led to better facilities and laboratories at the universities. It led to close links and cooperation between universities such as Harvard and Berkeley, and this would eventually lead to the development of ARPANET by DARPA.

Bush outlined his vision of an information management system called the *memex* (memory extender) in a famous essay *As We May Think* (Bush 1945). He envisaged the memex as a device electronically linked to a library that would be able to display books and films. It describes a proto-hypertext computer system and influenced the later development of hypertext systems.

Tim Berners-Lee (Fig. 57.2) is a British computer scientist and the inventor of the World Wide Web. He obtained a degree in physics from Oxford University in the mid-1970s, and he went to CERN for a short-term contract programming assignment in the early 1980s. CERN is a key European center for research in the nuclear field, and it employs several thousand physicists and scientists. One of the problems that scientists that CERN faced at that time was in keeping track of people, computers, documents, and databases. The center had many visiting scientists who spent several months there, as well as a large pool of permanent staff. There was no effective way at that time to share information among scientists.

A visiting scientist might need to obtain information or data from a CERN computer or to make the results of their research available to CERN. Berners-Lee

© Springer Nature Switzerland AG 2018
G. O'Regan, *The Innovation in Computing Companion*,
https://doi.org/10.1007/978-3-030-02619-6_57

Fig. 57.1 Vannevar Bush

Fig. 57.2 Tim Berners-
Lee. Creative Commons

developed a program called "Enquire"[1] to assist with information sharing and in keeping track of the work of visiting scientists. He returned to CERN in the mid-1980s to work on other projects, and he devoted part of his free time to consider solutions to the information-sharing problem. His solution was his invention of the *World Wide Web* in 1989.

He built on existing inventions such as the *Internet*, *hypertext*, and the *mouse*. Ted Nelson invented hypertext in the 1960s, and it allows links to be present in text. For example, a document such as a book contains a table of contents, an index, and

---

[1] The name "Enquire" came from a popular Victorian book called *Enquire Within Upon Everything* which was originally published in 1850.

a bibliography. These are all links to material that is either within the book itself or external to the book. The reader of a book can follow the link to obtain the internal or external information. Doug Engelbart invented the mouse in the 1960s (see Chap. 40), and it allows the cursor to be steered around the screen. The Internet (see Chap. 33) is the global system of interconnected computer networks (i.e., network of networks) that use the TCP/IP protocols to link electronic devices around the world.

The major leap that Berners-Lee made was essentially a marriage of the Internet, hypertext, and the mouse into what has become the World Wide Web. His vision (Berners-Lee 1999) and its subsequent realization benefited CERN and the wider world:

> Suppose that all information stored on computers everywhere were linked. Program computer to create a space where everything could be linked to everything.

The World Wide Web creates a space in which users can access information easily in any part of the world. This is done using only a web browser and web addresses. The user can then click on hyperlinks on web pages to access further relevant information that may be on an entirely different continent. Berners-Lee became the director of the World Wide Web Consortium, and this MIT-based organization sets the software standards for the web.

The invention of the World Wide Web was a revolution in computing. It transformed the Internet from mainly academic use to where it is now an integral part of peoples' lives. Users may now *surf the web*: i.e., hyperlink among the millions of computers in the world and obtain information easily. It is revolutionary in that:

- No single organization is controlling the web.
- No single computer is controlling the web.
- Millions of computers are interconnected.
- It is an enormous market place of billions of users.
- The web is not located in one physical location.
- The web is a space and not a physical thing.

Berners-Lee created a system that gives every web page a standard address called the Universal Resource Locator (URL). Each page is accessible via the hypertext transfer protocol (HTTP), and the page is formatted with the Hypertext Markup Language (HTML). Each page is visible using a web browser. The key features of his invention are described in Table 57.1.

Table 57.1 Features of World Wide Web

Feature	Description
URL	Universal Resource Identifier (later renamed to Universal Resource Locator (URL)) provides a unique address code for each web page
HTML	Hypertext Markup Language (HTML) is used for designing the layout of web pages
HTTP	The Hypertext Transport Protocol (HTTP) allows a new web page to be accessed from the current page
Browser	A browser is a client program that allows a user to interact with the pages and information on the World Wide Web

Berners-Lee invented the well-known terms such as URL, HTML, and World Wide Web. He wrote the first browser program that allowed users to access web pages throughout the world. Browsers are used to connect to remote computers over the Internet and to request, retrieve, and display the web pages on the local machine. The first web site was at CERN (*info.cern.ch*).

The early browsers included Gopher developed at the University of Minnesota and Mosaic developed at the University of Illinois. Netscape dominated the browser market until Microsoft developed Internet Explorer. The development of graphical browsers led to the commercialization of the World Wide Web.

The growth of the World Wide Web has been phenomenal, and exponential growth rate curves became a feature of newly formed Internet companies and their business plans (see Chap. 26). The World Wide Web has been applied to many areas including:

- Travel industry (booking flights, train tickets, and hotels)
- E-commerce (see Chap. 26)
- E-marketing
- Online shopping (e.g., www.amazon.com)
- Recruitment services (such as www.linkedin.com)
- Internet banking
- Newspapers and news channels
- Social media (e.g., Facebook)

Berners-Lee received the ACM Turing Award in 2017 for:

For inventing the World Wide Web, the first web browser, and the fundamental protocols and algorithms allowing the web to scale.

Chapter 58
Z3 and Z4 Computers

Konrad Zuse is considered *the father of the computer* in Germany (Fig. 58.1), as he built the world's first programmable machine (the Z3) in 1941. He was born in Berlin in 1910, and he studied civil engineering at the Technical University of Berlin.

He commenced working as a stress analyzer for Henschel after his graduation in 1935, and he resigned after a year with the intention of forming his own company to build automatic calculating machines. He commenced work on what would become the Z1 machine in 1936, and it employed the binary system, and metallic shafts could shift from position 0 to 1 and vice versa. The Z1 was operational by 1938.

He served briefly in the German Army, but Henschel helped Zuse to obtain a deferment from the army, as he was needed as an engineer. He remained affiliated to Henschel for the duration of the war, and he built the Z2 and Z3 machines there. The Z3 was operational in 1941, and it was the world's first programmable computer. Zuse started his own company in 1941, and this was the first company founded with the sole purpose of developing computers. The Z4 was almost complete as the Red Army advanced on Berlin in 1945, and Zuse left Berlin for Bavaria with the Z4 prior to the Russian advance.

Zuse designed the first high-level programming language between 1943 and 1945, and this language was called Plankalkül. He later re-started his company (Zuse KG), and he completed the Z4 in 1950. This was the first commercial computer as it was completed ahead of the Ferranti Mark 1, Univac, and LEO I computer. Its first customer was the Technical University of Zurich.

Zuse's results are especially impressive given that he was working alone in Germany, and he was unaware of the developments taking place in other countries. He independently implemented the principles of modern digital computers in isolation.

© Springer Nature Switzerland AG 2018
G. O'Regan, *The Innovation in Computing Companion*,
https://doi.org/10.1007/978-3-030-02619-6_58

Fig. 58.1 Konrad Zuse.
Courtesy of Horst Zuse,
Berlin

58.1   The Z1–Z3 Machines

Zuse commenced work on his first machine called the Z1 in 1936, and it was operational by 1938. It was demonstrated to a small number of people who saw it rattle and compute the determinant of a three by three matrix. It was essentially a binary electrically driven mechanical calculator with limited programmability. It could execute instructions read from the program punch cards, but the program itself was never loaded into the memory.

It employed the binary system and metallic shafts that could slide from position 0 to position 1 and vice versa. The machine was essentially a 22-bit *floating-point* value adder and subtracter. A decimal keyboard was used for input, and the output was decimal digits.

The machine included some control logic, which allowed it to perform more complex operations such as multiplications and division. These operations were performed by repeated additions for multiplication and repeated subtractions for division. The multiplication took approximately 5 s. The computer memory contained 64 22-bit words. Each word of memory could be read from and written to by the program punch cards and the control unit. It had a clock speed of 1 Hz and two floating-point registers of 22 bits each. The machine was unreliable and a reconstruction of it is in the Deutsches Technikmuseum in Berlin.

His next attempt was the creation of the Z2 machine, which aimed to improve on the Z1. This was a mechanical and relay *computer* created in *1939*. It used a similar mechanical *memory* but replaced the arithmetic and control logic with 600 electrical

Fig. 58.2 Zuse and the reconstructed Z3. Courtesy of Horst Zuse, Berlin

relay circuits. It used 16-bit fixed-point arithmetic instead of the 22-bit used in the Z1. It had a 16-bit word size, and the size of its memory was 64 words. It had a clock speed of 3 Hz.

The Z3 machine was built in 1941, and it was the first functional tape-stored-program-controlled computer (Fig. 58.2). It used 2600 telephone relays and binary number system and could perform floating-point arithmetic. It had a clock speed of 5 Hz and multiplication and division took 3 s. The input to the machine was with a decimal keyboard, and the output was on lamps that could display decimal numbers. The word length was 22 bits, and the size of the memory was 64 words.

It used a punched film for storing the sequence of program instructions. It could convert decimal to binary and back again. It was the first digital computer since it predates the Atanasoff-Berry Computer by 1 year. It was proven to be *Turing complete in 1998*. There is a reconstruction of the Z3 computer in the Deutsches Museum in Munich.

58.2 The Z4 Machine

The Z4 machine (Fig. 58.3) consisted of 2200 relays, a mechanical memory of 64 32-bit words, and a processor. The speed of the machine was approximately 1000 instructions per hour. The Henschel Aircraft Company had ordered the Z4 machine

Fig. 58.3 The Z4 Computer. Courtesy of Horst Zuse, Berlin

in 1942, but due to delays it was never actually delivered. The machine was almost complete by the end of the Second World War in 1945.

Konrad Zuse founded Zuse KG at Neukirchen (north of Frankfurt) in 1949. It was the first computer company in Germany, and the initial focus of the company was to restore and improve Zuse's Z4 machine, which had survived the Allied bombing of Berlin, and Zuse's subsequent move to Bavaria.

The Z4 was restored for the Institute of Applied Mathematics at the Eidgenössische Technische Hochschule (ETH) Zürich in Zurich. The restoration was complete in 1950, and it was delivered to the ETH later that year. It was one of the first operational computers in Europe at that time.

It was transferred to the French-German Research Institute of Saint-Louis in France in 1955, and today it is on display at the Deutsche Museum in Munich.

58.3   Plankalkül

The earliest high-level programming language was *Plankalkül* developed by Zuse in 1946. It means "Plan" and "Kalkül": i.e., a calculus of programs. The first Plankalkül program was run over 55 years after its conception, when the Free University of Berlin developed a compiler for the language in 2000.

It is a relatively modern language for a language designed in the mid-1940s, and it employs data structures and Boolean algebra, and more powerful data structures may be defined. Zuse showed that Plankalkül could be used to solve scientific and engineering problems, and he wrote several sample programs for sorting lists and searching a list for an element. The main features of Plankalkül are:

– It is a high-level language.
– Its fundamental data types are arrays and tuples of arrays.
– The while construct is used for iteration.
– Conditionals are addressed using guarded commands.
– There is no GOTO statement.
– Programs are non-recursive functions.
– The type of a variable is specified when it is used.

The main constructs of the language are variable assignment, arithmetical and logical operations, guarded commands, and while loops. There are also some list and set processing functions.

Chapter 59
Epilogue

This book has attempted to give a flavor of a selection of pivotal inventions that have shaped the computing field. It was not feasible to consider all inventions that merit inclusion, and the selection inevitably reflects the bias of the author.

We discussed a selection of historical inventions such as Babbage's Difference and Analytic Engine, Boole's symbolic logic, binary arithmetic as developed by Leibniz, and the von Neumann architecture which is the fundamental architecture underlying a computer.

We discussed a selection of historical analog and digital computers including Hollerith's tabulating machine which was developed in the late nineteenth century; Bush's differential analyzer which was developed in the 1920s; the Harvard Mark 1 analog computer; the Atanasoff-Berry computer (ABC); the Colossus computer which was developed at Bletchley Park in England; the ENIAC and EDVAC computers developed in the United States; the Manchester Mark 1 computer developed in England; the Z1–Z4 machines developed in Germany; and the LEO computers developed in England.

We discussed a selection of mainframes and minicomputers including the IBM System/360, the Amdahl 470 and 580 computers, and the DEC PDP-11 and VAX-11/780 minicomputers.

We discussed a selection of programming languages including Ada, which was developed by the US Military; C which was developed by Dennis Ritchie at Bell Labs; C++ which was developed by Bjarne Stroustrup at Bell Labs; COBOL which was developed by Grace Murray Hopper and the CODASYL committee; and Java which was developed by James Gosling at Sun Microsystems.

We discussed the invention of the transistor at Bell Labs, the invention of the integrated circuit at Texas Instruments, and the invention of the microprocessor at Intel. We discussed the development of operating systems such as Unix and MS/DOS.

We discussed a selection of home and personal computers including the MITS Altair, the Apple I and II computers, the Commodore PET and 64 computers, the IBM personal computer, and the Apple Macintosh. We discussed inventions related

© Springer Nature Switzerland AG 2018
G. O'Regan, *The Innovation in Computing Companion*,
https://doi.org/10.1007/978-3-030-02619-6_59

to personal computing including the mouse, the GUI, Atari video games, and Microsoft Office.

We discussed several inventions related to fixed line and mobile communications including the AXE system, mobile phones, Wi-Fi, and Iridium. We discussed the invention of the Internet and World Wide Web, which led to email, e-commerce, smartphones, and social media. We discussed the GPS technology.

We discussed a selection of innovations in the software engineering field including the Agile methodology, CMMI, formal methods, object-oriented design and development, open-source software development, software inspections, and software lifecycles.

We presented a selection of inventions related to commercial computing including databases and cloud computing; finally, we discussed miscellaneous inventions such as the ATM, AI, Eliza, digital photography, MP3 and digital music, robotics, and Wikipedia.

59.1 What Next in Computing?

The technological achievements in computing are extraordinary. The human race has embarked on an amazing journey from the development of tabulators in the late nineteenth century to the development of analog computers; to the development of the early bulky digital computers; to the development of early commercial computers; to the development of transistors and integrated circuits; to the IBM mainframes and digital minicomputers; to the development of the microprocessor and home computers; to the release of the IBM personal computer and the Apple Macintosh; to the rise of the Internet and World Wide Web; to the invention of the mobile phone; to the development of smartphones and social media; and so on.

We are living in a rapidly changing world where computer technology has driven innovation in almost all aspects of our world (e.g., communication, the media, medicine, automobiles, banking, and so on). These developments have led to major benefits to society, and it is natural to wonder where these innovations will lead. Can the extraordinary progress in the computing field continue given the limitations of Moore's law at the atomic level? Will robots one day perform much of the work done by humans? Will real progress be made in the AI field? Will it be possible someday for machines to achieve human-like intelligence? Will self-driving cars become a reality? Will technology assist in the elimination of poverty in the world? Will technology be used for good purposes? The pace of change is so relentless in that any predictions made are likely to be wide of the mark. All that we can say is that it will be interesting.

Test Yourself (Quiz 1)

1. Explain the significance of the ABC computer.
2. Explain the main features of Agile including scrum and sprints.
3. Explain the significance of the Apple Macintosh. What are Apple's key innovations?
4. Describe Gene Amdahl's contributions to IBM and the Amdahl Corporation.
5. Describe the components of the Analytic Engine, and explain their function.
6. Explain why Lady Ada Lovelace is considered the world's first programmer.
7. Discuss the significance of the Difference Engine.
8. Describe the Turing test. Is it an appropriate test of machine intelligence?
9. Describe Searle's Chinese room thought experiment, and discuss whether it is an effective rebuttal of machine intelligence.
10. Discuss the significance of Atari in the history of computing.
11. Explain binary numbers and their importance in the computing field.
12. Explain why Shannon's Master's thesis is a key milestone in computing.
13. Describe how Boolean logic may be employed in the design of digital circuits.
14. Explain the importance of Boole's equation $x^2 = x$ as a fundamental law of thought.
15. Discuss the similarities and differences between C and C++.
16. Explain cloud computing and distributed computing.
17. What is the CMMI?
18. Explain why Watt Humphrey is known as the father of software quality.
19. Describe Flower's work on the Colossus computer at Bletchley Park and how the machine was used to crack the Lorenz codes.
20. Describe Grace Murray Hopper's contribution to the COBOL programming language.
21. What is a compiler? Explain the various parts of a compiler.
22. What is a database? Explain the difference between a hierarchical, network, and relational database.
23. What is a relational database? What is an Oracle database?

© Springer Nature Switzerland AG 2018
G. O'Regan, *The Innovation in Computing Companion*,
https://doi.org/10.1007/978-3-030-02619-6

Test Yourself (Quiz 2)

1. Explain the significance of DEC's PDP-11 and VAX 11/780 minicomputers.
2. Explain the significance of the ENIAC and EDVAC computers.
3. Explain how Weizenbaum became a leading critic of AI, and explain his views on the ethics of AI.
4. Explain how electronic mail was invented.
5. Describe the business models employed in e-commerce.
6. What are formal methods and when should they be used?
7. What is GPS? Describe its applications.
8. What is a GUI?
9. Describe the Harvard Mark 1 and explain its significance.
10. Describe how Hollerith's tabulating machine led to the birth of IBM.
11. What is an integrated circuit and explain its significance?
12. Explain the significance of Moore's law.
13. What is the Internet and describe its history
14. What is the Iridium system? Why was the original system a commercial failure?
15. Discuss the importance of the Java programming language.
16. Describe Cerf's contributions to TCP/IP.
17. Explain the significance of LEO computers in the history of computing.
18. What is a microprocessor?
19. Explain why Gary Kildall has been described as the *man who could have been Bill Gates*.
20. Explain the significance of the Manchester "Baby" computer in the history of computing.
21. What is a mobile phone? Describe the history of the mobile phone.
22. What is a computer mouse? Describe its public demonstration at the "mother of all demos" in 1968.
23. Compare and contrast the achievements of Bill Gates and Steve Jobs.

© Springer Nature Switzerland AG 2018
G. O'Regan, *The Innovation in Computing Companion*,
https://doi.org/10.1007/978-3-030-02619-6

Test Yourself (Quiz 3)

1. What is a MP3 player?
2. What is MS/DOS? Describe the controversy with respect to the operating system for the original IBM PC.
3. Describe the main programs in Microsoft Office.
4. What is open-source software?
5. Explain the main parts of object-oriented development.
6. Explain the significance of home and personal computers.
7. Explain the significance of Don Estridge in the history of computing.
8. What errors did IBM make in the introduction of the IBM PC?
9. Describe Asimov's Laws of Robotics.
10. What is a smartphone?
11. Describe Fagan inspections. How effective are they in building quality into the software?
12. Describe the main software life cycles.
13. What is the significance of the IBM System/360 in the history of computing?
14. What is a transistor? Discuss Shockley's contributions to the computing field.
15. What is UNIX?
16. Describe the components of the von Neumann architecture, and explain each component.
17. Explain how Tim Berners-Lee invented the World Wide Web at CERN.
18. What is Wikipedia and how does it differ from a standard encyclopedia.
19. What is the World Wide Web?
20. What is Wi-Fi?
21. Discuss the significance of Zuse's Z3 and Z4 computers.
22. Describe the Plankalkül language, and explain why it took over 50 years for a Plankalkül program to be run.

© Springer Nature Switzerland AG 2018
G. O'Regan, *The Innovation in Computing Companion*,
https://doi.org/10.1007/978-3-030-02619-6

Glossary

ABC	Atanasoff-Berry computer
ACM	Association for Computing Machinery
ACS	Advanced Computer Systems
ADEC	Aiken Dahlgren electronic calculator
AI	Artificial intelligence
AMD	Advanced Micro Devices
AMP	Advanced Multimedia Product
AMPS	Advanced Mobile Phone System
ANS	Advanced Network Services
ANSI	American National Standards Institute
ARC	Augmentation Research Center
ARPA	Advanced Research Projects Agency
ASCC	Automatic Sequence Controlled Calculator
AT&T	American Telephone and Telegraph Company
ATM	Automated teller machine
ATMC	ATM controller
AXE	Automatic Exchange Electric
B2B	Business to business
B2C	Business to consumer
BASIC	Beginners All-purpose Symbolic Instruction Code
BCPL	Basic Combined Programming Language
BDS	BeiDou Navigation Satellite System
BIOS	Basic input/output system
BBN	Bolt, Beranek, and Newman
BTC	Bitcoin
C64	Commodore 64
CBA IPI	CMM-based appraisal internal process Improvement
CCD	Charge-coupled device
CCTV	Closed-circuit television

© Springer Nature Switzerland AG 2018
G. O'Regan, *The Innovation in Computing Companion*,
https://doi.org/10.1007/978-3-030-02619-6

CD	Compact disk
CDC	Control Data Corporation
CDMA	Code-division multiple access
CEO	Chief executive officer
CERN	Conseil Européen pour la Recherche Nucleaire
CERT	Computer emergency response team
CICS	Customer Information Control System
CMM®	Capability Maturity Model
CMMI®	Capability Maturity Model Integration
CMU	Carnegie Mellon University
COBOL	Common Business-Oriented Language
CODASYL	Conference/Committee on Data Systems Languages
CP/M	Control Program for Microcomputers
CPU	Central processing unit
CTR	Computing-Tabulating-Recording Company
DACS	De La Rue Automated Cash System
DARPA	Defense Advanced Research Project Agency
DB	Database
DBA	Database administrator
DBMS	Database management system
DCS	Digital camera system
DDL	Data definition language
DEC	Digital Equipment Corporation
DL	Data language
DML	Data manipulation language
DNS	Domain Name System
DoD	Department of Defence
DOS	Disk operating system
DRI	Digital Research Incorporated
DSDM	Dynamic systems development method
DSP	Digital signal processing
DVD	Digital versatile disk
EDSAC	Electronic delay storage automatic calculator
EDVAC	Electronic Discrete Variable Automatic Computer
EMCC	Eckert-Mauchly Computer Corporation
ENIAC	Electronic Numerical Integrator and Computer
ETH	Swiss Federal Institute of Technology, Zurich
ETSI	European Telecommunications Standards Institute
EULA	End-user license agreement
FDMA	Frequency division multiple access
FSF	Free Software Foundation
FTP	File Transfer Protocol
FOSS	Free and open-source software
GB	Gigabyte
GECOS	General Electric Comprehensive Operating System

GLONASS	Global Navigation Satellite System (Russia)
GM	General Motors
GNSS	Global Navigation Satellite System (Europe)
GNU	GNU's Not Unix
GPL	General Public License
GPRS	General Packet Radio Services
GPS	Global Positioning System
GSM	Global System for Mobile Communication
GUI	Graphical user interface
HCI	Human-computer interaction
HOLWG	Higher-Order Language Working Group
HMD	Head-mounted display
HP	Hewlett-Packard
HTML	Hypertext Markup Language
HTTP	Hypertext Transport Protocol
IaaS	Infrastructure as a Service
IBM	International Business Machines
IC	Integrated circuit
ICL	International Computers Ltd.
IDMS	Integrated Database Management System
IDS	Integrated Data Store
IEC	International Electrotechnical Commission
IEEE	Institute of Electrical and Electronic Engineers
IMAP	Internet Message Application Protocol
IMP	Interface Message Processor
IMS	Information management system
*i***OS**	*i*Phone operating system
IP	Internet Protocol
IPO	Initial public offering
ISO	International Standards Organization
JAD	Joint Application Development
JCP	Java Community Process
JDK	Java Development Kit
JIT	Just-in-time
JPEG	Joint Photographic Experts Group
JVM	Java virtual machine
KLOC	Thousand Lines of Code
LAN	Local area network
LCD	Liquid-crystal display
LED	Light-emitting diode
LEO	Lyons Electronic Office
LSI	Large-scale integration
MADC	Manchester Automatic Digital Computer
MEO	Medium Earth orbit
MIDI	Musical Instrument Digital Interface

MIPS	Million instructions per second
MIT	Massachusetts Institute of Technology
MITS	Micro Instrumentation and Telemetry System
MP	Megapixel
MPEG	Movie Picture Experts Group
MS/DOS	Microsoft Disk Operating System
MSI	Medium-scale integration
MTX	Mobile telephone exchange
NAP	Network Access Point
NASA	National Aeronautics and Space Administration
NBS	National Bureau of Standards
NCP	Network Control Protocol
NCR	National Cash Register
NLS	oN-Line System
NMT	Nordic Mobile Telephone system
NR	Norwegian Research
NSF	National Science Foundation
OBE	Order of the British Empire
OOD	Object-oriented design
OS	Operating system
OSS	Open-source software
PaaS	Platform as a Service
PARC	Palo Alto Research Center
PC	Personal computer
P-CMM	People Capability Maturity Model
PDA	Personal digital assistant
PDP	Programmed Data Processor
PET	Personal Electronic Transactor
PIN	Personal identification number
PL/M	Programming Language for Microcomputers
POP	Post Office Protocol
PSP	Personal Software Process
RAD	Rapid application development
RAM	Random access memory
RDBMS	Relational database management system
RCA	Root cause analysis
ROM	Read-only memory
RUP	Rational Unified Process
SaaS	Software as a Service
SAGE	Semiautomatic Ground Environment
SEI	Software Engineering Institute
SID	Sound Interface Device
SILK	Speech, Image, Language, Knowledge
SIM	Subscriber identity modules
SLIP	Symmetric List Processor

SMS	Short message service
SMTP	Simple Mail Transfer Protocol
SOA	Service-oriented architecture
SPREAD	System Programming Research Engineering and Development
SQL	Structured Query Language
SRI	Stanford Research Institute
SSEC	Selective Sequence Electronic Computer
SSEC	Small-scale experimental computer
SSI	Small-scale integration
TACs	Total Access Communication
TCP	Transmission Control Protocol
TI	Texas Instruments
TSP	Team software process
UCD	User-centered design
UCLA	University of California (Los Angeles)
UDP	User Datagram Protocol
ULSI	Ultra-large-scale integration
UML	Unified Modeling Language
UNIVAC	Universal Automatic Computer
URL	Universal Resource Locator
UTC	Coordinated Universal Time
VAX	Virtual Address eXtension
VBA	Visual Basic for Applications
VCS	Video Computer System
VDM	Vienna Development Method
VLSI	Very-large-scale integration
VMS	Virtual Memory System
VoIP	Voice over Internet Protocol
VUI	Voice user interface
WCDMA	Wideband CDMA
WIMP	Windows, icons, menus, and pointers
WLAN	Wireless LAN
WPA	Wi-Fi Protected Area
WYSIWYG	What you see is what you get
WWW	World Wide Web
XP	Extreme Programming

References

Ada, Augusta, Countess of Lovelace (1842) Sketch of the Analytic Engine invented by Charles Babbage. L.F. Menabrea, Bibliothèque Universelle de Genève, October, 1842, No. 82 Translated by Ada, Augusta, Countess of Lovelace

Aho A, Ullman J (1977) Principles of compiler design. Addison Wesley, Reading

Arnold K, Gosling J, Holmes D (2013) The Java programming language, 5th edn. Prentice Hall, Upper Saddle River

Bagnall B (2012) Commodore. A company on the edge, 2nd edn. Variant Press, Winnipeg

BBC Magazine (2017) Can we teach robots ethics. 17 October 2017

Beck K et al (2001) Manifesto for agile software development. Agile Alliance. http://agilemanifesto.org/

Berners-Lee T (1999) Weaving the web. Harper. 1999

Bloomberg Business Week Magazine (2004) The man who could have been Bill Gates. October 2004

Boehm B (1981) Software engineering economics. Prentice Hall, Upper Saddle River

Boehm B (1988) A spiral model for software development and enhancement. Computer 21(5):61–72

Boole G (1848) The calculus of logic. Cambridge and Dublin Math J III:183–198

Boole G (1958) An investigation into the laws of thought. Dover Publications, New York. (First published in 1854)

Boyer C (2004) The 360 revolution. IBM ftp://ftp.software.ibm.com/s390/misc/bookoffer/download/360revolution_040704.pdf

Brooks F (1975) The Mythical man month. Addison Wesley, Boston

Brown MJD (1990) Rationale for the development of the UK defence standards for safety critical computer software. Proc. COMPASS '90, Washington DC, USA, June 1990

Bush V (1945) As we may think. The Atlantic Monthly 176(1):101–108

Chaum D (1982) Blind signatures for untraceable payments. Adv Cryptol Proc Crypto 82(3):199–203

Chrissis MB, Conrad M, Shrum S (2011) CMMI. Guidelines for process integration and product improvement, SEI series in software engineering, 3rd edn. Addison Wesley, Boston

Codd EF (1970) A relational model of data for large shared data banks. Commun ACM 13(6):377–387

Date CJ (1981) An introduction to database systems, The systems programming series, 3rd edn. Addison Wesley, Reading

Deitel HM (1990) Operating systems, 2nd edn. Addison Wesley, Reading

Edwards B (2011) The history of Atari computers. PC World. April 21st 2011. http://www.vintage-computing.com/index.php/archives/760/the-history-of-atari-computers

Fagan M (1976) Design and code inspections to reduce errors in software development. IBM Syst J 15(3):182–210

Ferry G (2003) A computer called LEO: lyons tea shops and the world's first office computer. Fourth Estate Ltd, London

Gilb T, Graham D (1994) Software inspections. Addison Wesley, Wokingham/Reading

Gosling J et al (2014) The Java language specification. Addison Wesley Professional, Reading

Greenfield A (2017) Rise of the machines. Who is the internet of things good for? Guardian Article. https://www.theguardian.com/technology/2017/jun/06/internet-of-things-smart-home-smart-city. 6 June 2017

Halfhill T (1994) R.I.P. Commodore. 1954–1994. A look at an innovative industry pioneer, whose achievements have been largely forgotten. Byte Magazine, August 1994

Hinchey M, Bowen J (eds) (1995) Applications of formal methods, Prentice Hall international series in computer science. Prentice Hall, London/New York

Humphry W (1989) Managing the software process. Addison Wesley, Boston

IGN presents the history of Atari (2014) http://www.ign.com/articles/2014/03/20/ign-presents-the-history-of-atari

Isaacson W (2011) Steve Jobs. Little, Brown Press, London

Jacaobson I et al (1999) The unified software development process. Addison Wesley, Reading

Kahn B, Cerf V (1974) Protocol for packet network interconnections. IEEE Trans Commun Technol 22(5):637–648

Kernighan B, Ritchie D (1978) The C programming language, 1st edn. Prentice Hall Software Series, Englewood Cliffs

Leibniz WG (1703) *Explication de l'Arithmétique Binaire Memoires de l'Academie Royale des Sciences*

Licklider JCR (March 1960) Man-computer symbiosis. IRE Transactions on Human Factors in Electronics HFE 1:4–11

Lada Ada Lovelace (1842) Sketch of the Analytic Engine. Invented by Charles Babbage. L.F. Menabrea. Bibliothèque Universelle de Genève. Translated by Lada Ada Lovelace. 1842

Malmsten E, Portanger E (2002) Boo hoo. $135 million, 18 months. A dot.com story from concept to catastrophe. Arrow, London

Malone M (2014) The intel trilogy. How Robert Noyce, Gordon Moore, and Andy Grove Built the World's Most Important Company. Harper Collins Publishers. 2014

McHale D (1985) Boole. Cork University Press, Cork

McKay S (2011) The secret life of Bletchley Park. Aurum Press, London

Meurling J et al (2001) The Ericsson Chronicles. 125 years in telecommunications. Informationsforlaget, Sweden

Misra P, Enge P (2010) Global positioning system, 2nd edn. Ganga-Jamuna Press, Lincoln

Moore G (April 19,1965) Cramming more components onto integrated circuits. Electron Mag, 38(8)

Munson GE (2011) The rise and fall of unimation inc. A story of robotics innovation and triumph that changed the world. Robot Magazine 24:36

Nakamoto S (2008) Bitcoin: a peer-to-peer electronic cash system. https://bitcoin.org/bitcoin.pdf. 31st October 2008

O'Regan G (2010) Introduction to software process improvement. Springer, New York

O'Regan G (2013a) Mathematics in computing. Springer, London

O'Regan G (2013b) Giants of computing. Springer, Heidelberg

O'Regan G (2014) Introduction to software quality. Springer, Cham

O'Regan G (2015) Pillars of computing. Springer, Cham

O'Regan G (2017a) Concise guide to software engineering. Springer, Cham

O'Regan G (2017b) Concise guide to formal methods. Springer, Cham

Parnas D (1972) On the criteria to be used in decomposing systems into modules. Commun ACM 15(12):1053–1058

Pugh EW (2009) Building IBM: shaping an industry and its technology. MIT Press, Cambridge, MA

Robbins A (2005) Unix in a Nutshell, 4th edn. O'Reilly Media, Sebastopol

Royce W (1970) Managing the development of large software systems in: Technical papers of Western Electronic Show and Convention (WesCon) August 25–28, 1970, Los Angeles, USA

Schaefer MW (2014) The tao of twitter. Changing your life and business 140 characters at a time, 2nd edn. McGraw-Hill, New York

Schein E et al (2004) DEC is dead: long live DEC. Berrett-Koehler Publishers, San Francisco

Searle J (1980) Minds, brains, and programs. Behav Brain Sci 3:417–457

Claude Shannon (1937) A symbolic analysis of relay and switching circuits. Masters thesis. Massachusetts Institute of Technology

Shockley W (1950) Electrons and holes in semiconductors with applications to transistor electronics. Van Nostrand, New York

Software Engineering Institute (2006) CMMI executive overview. Presentation by the SEI

Sommerville I (2011) Software engineering, 9th edn. Pearson, Boston

Stallman RM (2002) Free software, free society, 2nd edn. Free Software Foundation, Inc, Boston

Stern N et al (2005) Cobol for 21st century, 11th edn. Wiley, New York

Stroustrup B (2013) C++ programming language, 4th edn. Addison Wesley, Upper Saddle River

Turing A (1950) Computing, machinery and intelligence. Mind 49:433–460

von Neumann J (1945) First draft of a report on the EDVAC. University of Pennsylvania, Philadelphia

Weisfield M (2013) The object-oriented thought process, 4th edn. Addison-Wesley Professional, Indianapolis

Weizenbaum J (1966) ELIZA. A computer program for the study of natural language communication between man and machine. Commun ACM 9(1):36–45

Weizenbaum J (1976) Computer power and human reason: from judgments to calculation. W.H. Freeman & Co Ltd, San Francisco

Index

© Springer Nature Switzerland AG 2018
G. O'Regan, *The Innovation in Computing Companion*,
https://doi.org/10.1007/978-3-030-02619-6